Deep Learning:
Methods and Applications

Deep Learning: Methods and Applications

Li Deng

Microsoft Research
One Microsoft Way
Redmond, WA 98052; USA
deng@microsoft.com

Dong Yu

Microsoft Research
One Microsoft Way
Redmond, WA 98052; USA
Dong.Yu@microsoft.com

the essence of knowledge

Boston — Delft

Foundations and Trends® in Signal Processing

Published, sold and distributed by:
now Publishers Inc.
PO Box 1024
Hanover, MA 02339
United States
Tel. +1-781-985-4510
www.nowpublishers.com
sales@nowpublishers.com

Outside North America:
now Publishers Inc.
PO Box 179
2600 AD Delft
The Netherlands
Tel. +31-6-51115274

The preferred citation for this publication is

L. Deng and D. Yu. *Deep Learning: Methods and Applications.* Foundations and Trends® in Signal Processing, vol. 7, nos. 3–4, pp. 197–387, 2013.

This *Foundations and Trends*® issue was typeset in LaTeX using a class file designed by Neal Parikh. Printed on acid-free paper.

ISBN: 978-1-60198-814-0
© 2014 L. Deng and D. Yu

Foundations and Trends® in Signal Processing
Volume 7, Issues 3–4, 2013
Editorial Board

Editorial Scope

Topics

Foundations and Trends® in Signal Processing publishes survey and tutorial articles in the following topics:

- Adaptive signal processing
- Audio signal processing
- Biological and biomedical signal processing
- Complexity in signal processing
- Digital signal processing
- Distributed and network signal processing
- Image and video processing
- Linear and nonlinear filtering
- Multidimensional signal processing
- Multimodal signal processing
- Multirate signal processing
- Multiresolution signal processing
- Nonlinear signal processing
- Randomized algorithms in signal processing
- Sensor and multiple source signal processing, source separation
- Signal decompositions, subband and transform methods, sparse representations
- Signal processing for communications
- Signal processing for security and forensic analysis, biometric signal processing
- Signal quantization, sampling, analog-to-digital conversion, coding and compression
- Signal reconstruction, digital-to-analog conversion, enhancement, decoding and inverse problems
- Speech/audio/image/video compression
- Speech and spoken language processing
- Statistical/machine learning
- Statistical signal processing

Information for Librarians

Foundations and Trends® in Signal Processing, 2013, Volume 7, 4 issues. ISSN paper version 1932-8346. ISSN online version 1932-8354. Also available as a combined paper and online subscription.

Foundations and Trends® in Signal Processing
Vol. 7, Nos. 3–4 (2013) 197–387
© 2014 L. Deng and D. Yu
DOI: 10.1561/2000000039

Deep Learning: Methods and Applications

Li Deng
Microsoft Research
One Microsoft Way
Redmond, WA 98052; USA
deng@microsoft.com

Dong Yu
Microsoft Research
One Microsoft Way
Redmond, WA 98052; USA
Dong.Yu@microsoft.com

Contents

Endorsement

"In the past few years, deep learning has rapidly evolved into the de-facto approach for acoustic modeling in automatic speech recognition (ASR), showing tremendous improvement in accuracy, robustness, and cross-language generalizability over conventional approaches. This timely book is written by the pioneers of deep learning innovations and applications to ASR, who, as early as 2010, first succeeded in large vocabulary speech recognition using deep learning. This was accomplished using a special form of the deep neural net, developed by the authors, perfectly fit for fast decoding as required by industrial deployment of ASR technology. In addition to recounting this remarkable advance which ignited the industry-scale adoption of deep learning in ASR, this book also provides an overview of a sweeping range of up-to-date deep learning methodologies and its application to a variety of signal and information processing tasks, including not only ASR but also computer vision, language modeling, text processing, multimodal learning, and information retrieval. This is the first and the most valuable book for "deep and wide learning" of deep learning, not to be missed by anyone who wants to know the breath taking impact of deep learning in many facets of information processing, especially ASR, all of vital importance to our modern technological society."

— Sadaoki Furui, President of Toyota Technological Institute at Chicago, and Professor at the Tokyo Institute of Technology

Abstract

This monograph provides an overview of general deep learning methodology and its applications to a variety of signal and information processing tasks. The application areas are chosen with the following three criteria in mind: (1) expertise or knowledge of the authors; (2) the application areas that have already been transformed by the successful use of deep learning technology, such as speech recognition and computer vision; and (3) the application areas that have the potential to be impacted significantly by deep learning and that have been experiencing research growth, including natural language and text processing, information retrieval, and multimodal information processing empowered by multi-task deep learning.

L. Deng and D. Yu. *Deep Learning: Methods and Applications*. Foundations and Trends® in Signal Processing, vol. 7, nos. 3–4, pp. 197–387, 2013.
DOI: 10.1561/2000000039.

1

Introduction

1.1 Definitions and background

Since 2006, deep structured learning, or more commonly called deep learning or hierarchical learning, has emerged as a new area of machine learning research [20, 163]. During the past several years, the techniques developed from deep learning research have already been impacting a wide range of signal and information processing work within the traditional and the new, widened scopes including key aspects of machine learning and artificial intelligence; see overview articles in [7, 20, 24, 77, 94, 161, 412], and also the media coverage of this progress in [6, 237]. A series of workshops, tutorials, and special issues or conference special sessions in recent years have been devoted exclusively to deep learning and its applications to various signal and information processing areas. These include:

- 2008 NIPS Deep Learning Workshop;

- 2009 NIPS Workshop on Deep Learning for Speech Recognition and Related Applications;

- 2009 ICML Workshop on Learning Feature Hierarchies;

- 2011 ICML Workshop on Learning Architectures, Representations, and Optimization for Speech and Visual Information Processing;

- 2012 ICASSP Tutorial on Deep Learning for Signal and Information Processing;

- 2012 ICML Workshop on Representation Learning;

- 2012 Special Section on Deep Learning for Speech and Language Processing in IEEE Transactions on Audio, Speech, and Language Processing (T-ASLP, January);

- 2010, 2011, and 2012 NIPS Workshops on Deep Learning and Unsupervised Feature Learning;

- 2013 NIPS Workshops on Deep Learning and on Output Representation Learning;

- 2013 Special Issue on Learning Deep Architectures in IEEE Transactions on Pattern Analysis and Machine Intelligence (T-PAMI, September).

- 2013 International Conference on Learning Representations;

- 2013 ICML Workshop on Representation Learning Challenges;

- 2013 ICML Workshop on Deep Learning for Audio, Speech, and Language Processing;

- 2013 ICASSP Special Session on New Types of Deep Neural Network Learning for Speech Recognition and Related Applications.

The authors have been actively involved in deep learning research and in organizing or providing several of the above events, tutorials, and editorials. In particular, they gave tutorials and invited lectures on this topic at various places. Part of this monograph is based on their tutorials and lecture material.

Before embarking on describing details of deep learning, let's provide necessary definitions. Deep learning has various closely related definitions or high-level descriptions:

- **Definition 1**: A class of machine learning techniques that exploit many layers of non-linear information processing for supervised or unsupervised feature extraction and transformation, and for pattern analysis and classification.

- ***Definition 2***: "A sub-field within machine learning that is based on algorithms for learning multiple levels of representation in order to model complex relationships among data. Higher-level features and concepts are thus defined in terms of lower-level ones, and such a hierarchy of features is called a deep architecture. Most of these models are based on unsupervised learning of representations." (Wikipedia on "Deep Learning" around March 2012.)

- ***Definition 3***: "A sub-field of machine learning that is based on learning several levels of representations, corresponding to a hierarchy of features or factors or concepts, where higher-level concepts are defined from lower-level ones, and the same lower-level concepts can help to define many higher-level concepts. Deep learning is part of a broader family of machine learning methods based on learning representations. An observation (e.g., an image) can be represented in many ways (e.g., a vector of pixels), but some representations make it easier to learn tasks of interest (e.g., is this the image of a human face?) from examples, and research in this area attempts to define what makes better representations and how to learn them." (Wikipedia on "Deep Learning" around February 2013.)

- ***Definition 4***: "Deep learning is a set of algorithms in machine learning that attempt to learn in multiple levels, corresponding to different levels of abstraction. It typically uses artificial neural networks. The levels in these learned statistical models correspond to distinct levels of concepts, where higher-level concepts are defined from lower-level ones, and the same lower-level concepts can help to define many higher-level concepts." See Wikipedia http://en.wikipedia.org/wiki/Deep_learning on "Deep Learning" as of this most recent update in October 2013.

- ***Definition 5***: "Deep Learning is a new area of Machine Learning research, which has been introduced with the objective of moving Machine Learning closer to one of its original goals: Artificial

Intelligence. Deep Learning is about learning multiple levels of representation and abstraction that help to make sense of data such as images, sound, and text." See https://github.com/lisa-lab/DeepLearningTutorials

Note that the deep learning that we discuss in this monograph is about learning with deep architectures for signal and information processing. It is not about deep understanding of the signal or information, although in many cases they may be related. It should also be distinguished from the overloaded term in educational psychology: "Deep learning describes an approach to learning that is characterized by active engagement, intrinsic motivation, and a personal search for meaning." http://www.blackwellreference.com/public/tocnode?id=g9781405161251_chunk_g97814051612516_ss1-1

Common among the various high-level descriptions of deep learning above are two key aspects: (1) models consisting of multiple layers or stages of nonlinear information processing; and (2) methods for supervised or unsupervised learning of feature representation at successively higher, more abstract layers. Deep learning is in the intersections among the research areas of neural networks, artificial intelligence, graphical modeling, optimization, pattern recognition, and signal processing. Three important reasons for the popularity of deep learning today are the drastically increased chip processing abilities (e.g., general-purpose graphical processing units or GPGPUs), the significantly increased size of data used for training, and the recent advances in machine learning and signal/information processing research. These advances have enabled the deep learning methods to effectively exploit complex, compositional nonlinear functions, to learn distributed and hierarchical feature representations, and to make effective use of both labeled and unlabeled data.

Active researchers in this area include those at University of Toronto, New York University, University of Montreal, Stanford University, Microsoft Research (since 2009), Google (since about 2011), IBM Research (since about 2011), Baidu (since 2012), Facebook (since 2013), UC-Berkeley, UC-Irvine, IDIAP, IDSIA, University College London, University of Michigan, Massachusetts Institute of

Technology, University of Washington, and numerous other places; see http://deeplearning.net/deep-learning-research-groups-and-labs/ for a more detailed list. These researchers have demonstrated empirical successes of deep learning in diverse applications of computer vision, phonetic recognition, voice search, conversational speech recognition, speech and image feature coding, semantic utterance classification, natural language understanding, hand-writing recognition, audio processing, information retrieval, robotics, and even in the analysis of molecules that may lead to discovery of new drugs as reported recently by [237].

In addition to the reference list provided at the end of this monograph, which may be outdated not long after the publication of this monograph, there are a number of excellent and frequently updated reading lists, tutorials, software, and video lectures online at:

- http://deeplearning.net/reading-list/
- http://ufldl.stanford.edu/wiki/index.php/
  UFLDL_Recommended_Readings
- http://www.cs.toronto.edu/~hinton/
- http://deeplearning.net/tutorial/
- http://ufldl.stanford.edu/wiki/index.php/UFLDL_Tutorial

1.2 Organization of this monograph

The rest of the monograph is organized as follows:

In Section 2, we provide a brief historical account of deep learning, mainly from the perspective of how speech recognition technology has been hugely impacted by deep learning, and how the revolution got started and has gained and sustained immense momentum.

In Section 3, a three-way categorization scheme for a majority of the work in deep learning is developed. They include unsupervised, supervised, and hybrid deep learning networks, where in the latter category unsupervised learning (or pre-training) is exploited to assist the subsequent stage of supervised learning when the final tasks pertain to classification. The supervised and hybrid deep networks often have the

same type of architectures or the structures in the deep networks, but
the unsupervised deep networks tend to have different architectures
from the others.

Sections 4–6 are devoted, respectively, to three popular types of
deep architectures, one from each of the classes in the three-way categorization scheme reviewed in Section 3. In Section 4, we discuss
in detail deep autoencoders as a prominent example of the unsupervised deep learning networks. No class labels are used in the learning,
although supervised learning methods such as back-propagation are
cleverly exploited when the input signal itself, instead of any label
information of interest to possible classification tasks, is treated as the
"supervision" signal.

In Section 5, as a major example in the hybrid deep network category, we present in detail the deep neural networks with unsupervised
and largely generative pre-training to boost the effectiveness of supervised training. This benefit is found critical when the training data
are limited and no other appropriate regularization approaches (i.e.,
dropout) are exploited. The particular pre-training method based on
restricted Boltzmann machines and the related deep belief networks
described in this section has been historically significant as it ignited
the intense interest in the early applications of deep learning to speech
recognition and other information processing tasks. In addition to this
retrospective review, subsequent development and different paths from
the more recent perspective are discussed.

In Section 6, the basic deep stacking networks and their several
extensions are discussed in detail, which exemplify the discriminative, supervised deep learning networks in the three-way classification
scheme. This group of deep networks operate in many ways that are
distinct from the deep neural networks. Most notably, they use target
labels in constructing *each* of many layers or modules in the overall
deep networks. Assumptions made about part of the networks, such as
linear output units in each of the modules, simplify the learning algorithms and enable a much wider variety of network architectures to
be constructed and learned than the networks discussed in Sections 4
and 5.

In Sections 7–11, we select a set of typical and successful applications of deep learning in diverse areas of signal and information processing. In Section 7, we review the applications of deep learning to speech recognition, speech synthesis, and audio processing. Subsections surrounding the main subject of speech recognition are created based on several prominent themes on the topic in the literature.

In Section 8, we present recent results of applying deep learning to language modeling and natural language processing, where we highlight the key recent development in embedding symbolic entities such as words into low-dimensional, continuous-valued vectors.

Section 9 is devoted to selected applications of deep learning to information retrieval including web search.

In Section 10, we cover selected applications of deep learning to image object recognition in computer vision. The section is divided to two main classes of deep learning approaches: (1) unsupervised feature learning, and (2) supervised learning for end-to-end and joint feature learning and classification.

Selected applications to multi-modal processing and multi-task learning are reviewed in Section 11, divided into three categories according to the nature of the multi-modal data as inputs to the deep learning systems. For single-modality data of speech, text, or image, a number of recent multi-task learning studies based on deep learning methods are reviewed in the literature.

Finally, conclusions are given in Section 12 to summarize the monograph and to discuss future challenges and directions.

This short monograph contains the material expanded from two tutorials that the authors gave, one at APSIPA in October 2011 and the other at ICASSP in March 2012. Substantial updates have been made based on the literature up to January 2014 (including the materials presented at NIPS-2013 and at IEEE-ASRU-2013 both held in December of 2013), focusing on practical aspects in the fast development of deep learning research and technology during the interim years.

2

Some Historical Context of Deep Learning

Until recently, most machine learning and signal processing techniques had exploited shallow-structured architectures. These architectures typically contain at most one or two layers of nonlinear feature transformations. Examples of the shallow architectures are Gaussian mixture models (GMMs), linear or nonlinear dynamical systems, conditional random fields (CRFs), maximum entropy (MaxEnt) models, support vector machines (SVMs), logistic regression, kernel regression, multilayer perceptrons (MLPs) with a single hidden layer including extreme learning machines (ELMs). For instance, SVMs use a shallow linear pattern separation model with one or zero feature transformation layer when the kernel trick is used or otherwise. (Notable exceptions are the recent kernel methods that have been inspired by and integrated with deep learning; e.g. [9, 53, 102, 377]). Shallow architectures have been shown effective in solving many simple or well-constrained problems, but their limited modeling and representational power can cause difficulties when dealing with more complicated real-world applications involving natural signals such as human speech, natural sound and language, and natural image and visual scenes.

Human information processing mechanisms (e.g., vision and audition), however, suggest the need of deep architectures for extracting complex structure and building internal representation from rich sensory inputs. For example, human speech production and perception systems are both equipped with clearly layered hierarchical structures in transforming the information from the waveform level to the linguistic level [11, 12, 74, 75]. In a similar vein, the human visual system is also hierarchical in nature, mostly in the perception side but interestingly also in the "generation" side [43, 126, 287]). It is natural to believe that the state-of-the-art can be advanced in processing these types of natural signals if efficient and effective deep learning algorithms can be developed.

Historically, the concept of deep learning originated from artificial neural network research. (Hence, one may occasionally hear the discussion of "new-generation neural networks.") Feed-forward neural networks or MLPs with many hidden layers, which are often referred to as deep neural networks (DNNs), are good examples of the models with a deep architecture. Back-propagation (BP), popularized in 1980s, has been a well-known algorithm for learning the parameters of these networks. Unfortunately BP alone did not work well in practice then for learning networks with more than a small number of hidden layers (see a review and analysis in [20, 129]. The pervasive presence of local optima and other optimization challenges in the non-convex objective function of the deep networks are the main source of difficulties in the learning. BP is based on local gradient information, and starts usually at some random initial points. It often gets trapped in poor local optima when the batch-mode or even stochastic gradient descent BP algorithm is used. The severity increases significantly as the depth of the networks increases. This difficulty is partially responsible for steering away most of the machine learning and signal processing research from neural networks to shallow models that have convex loss functions (e.g., SVMs, CRFs, and MaxEnt models), for which the global optimum can be efficiently obtained at the cost of reduced modeling power, although there had been continuing work on neural networks with limited scale and impact (e.g., [42, 45, 87, 168, 212, 263, 304].

The optimization difficulty associated with the deep models was empirically alleviated when a reasonably efficient, unsupervised learning algorithm was introduced in the two seminar papers [163, 164]. In these papers, a class of deep generative models, called deep belief network (DBN), was introduced. A DBN is composed of a stack of restricted Boltzmann machines (RBMs). A core component of the DBN is a greedy, layer-by-layer learning algorithm which optimizes DBN weights at time complexity linear to the size and depth of the networks. Separately and with some surprise, initializing the weights of an MLP with a correspondingly configured DBN often produces much better results than that with the random weights. As such, MLPs with many hidden layers, or deep neural networks (DNN), which are learned with unsupervised DBN pre-training followed by back-propagation fine-tuning is sometimes also called DBNs in the literature [67, 260, 258]. More recently, researchers have been more careful in distinguishing DNNs from DBNs [68, 161], and when DBN is used to initialize the training of a DNN, the resulting network is sometimes called the DBN–DNN [161].

Independently of the RBM development, in 2006 two alternative, non-probabilistic, non-generative, unsupervised deep models were published. One is an autoencoder variant with greedy layer-wise training much like the DBN training [28]. Another is an energy-based model with unsupervised learning of sparse over-complete representations [297]. They both can be effectively used to pre-train a deep neural network, much like the DBN.

In addition to the supply of good initialization points, the DBN comes with other attractive properties. First, the learning algorithm makes effective use of unlabeled data. Second, it can be interpreted as a probabilistic generative model. Third, the over-fitting problem, which is often observed in the models with millions of parameters such as DBNs, and the under-fitting problem, which occurs often in deep networks, can be effectively alleviated by the generative pre-training step. An insightful analysis on what kinds of speech information DBNs can capture is provided in [259].

Using hidden layers with many neurons in a DNN significantly improves the modeling power of the DNN and creates many closely optimal configurations. Even if parameter learning is trapped into a local optimum, the resulting DNN can still perform quite well since the chance of having a poor local optimum is lower than when a small number of neurons are used in the network. Using deep and wide neural networks, however, would cast great demand to the computational power during the training process and this is one of the reasons why it is not until recent years that researchers have started exploring both deep and wide neural networks in a serious manner.

Better learning algorithms and different nonlinearities also contributed to the success of DNNs. Stochastic gradient descend (SGD) algorithms are the most efficient algorithm when the training set is large and redundant as is the case for most applications [39]. Recently, SGD is shown to be effective for parallelizing over many machines with an asynchronous mode [69] or over multiple GPUs through pipelined BP [49]. Further, SGD can often allow the training to jump out of local optima due to the noisy gradients estimated from a single or a small batch of samples. Other learning algorithms such as Hessian free [195, 238] or Krylov subspace methods [378] have shown a similar ability.

For the highly non-convex optimization problem of DNN learning, it is obvious that better parameter initialization techniques will lead to better models since optimization starts from these initial models. What was not obvious, however, is how to efficiently and effectively initialize DNN parameters and how the use of large amounts of training data can alleviate the learning problem until more recently [28, 20, 100, 64, 68, 163, 164, 161, 323, 376, 414]. The DNN parameter initialization technique that attracted the most attention is the unsupervised pretraining technique proposed in [163, 164] discussed earlier.

The DBN pretraining procedure is not the only one that allows effective initialization of DNNs. An alternative unsupervised approach that performs equally well is to pretrain DNNs layer by layer by considering each pair of layers as a de-noising autoencoder regularized by setting a random subset of the input nodes to zero [20, 376]. Another alternative is to use *contractive* autoencoders for the same purpose by favoring representations that are more robust to the input variations, i.e., penalizing the gradient of the activities of the hidden units with respect to the inputs [303]. Further, Ranzato et al. [294] developed the sparse encoding symmetric machine (SESM), which has a very similar architecture to RBMs as building blocks of a DBN. The SESM may also be used to effectively initialize the DNN training. In addition to unsupervised pretraining using greedy layer-wise procedures [28, 164, 295], the supervised pretraining, or sometimes called discriminative pretraining, has also been shown to be effective [28, 161, 324, 432] and in cases where labeled training data are abundant performs better than the unsupervised pretraining techniques. The idea of the discriminative pretraining is to start from a one-hidden-layer MLP trained with the BP algorithm. Every time when we want to add a new hidden layer we replace the output layer with a randomly initialized new hidden and output layer and train the whole new MLP (or DNN) using the BP algorithm. Different from the unsupervised pretraining techniques, the discriminative pretraining technique requires labels.

Researchers who apply deep learning to speech and vision analyzed what DNNs capture in speech and images. For example, [259] applied a dimensionality reduction method to visualize the relationship among the feature vectors learned by the DNN. They found that the DNN's hidden activity vectors preserve the similarity structure of the feature vectors at multiple scales, and that this is especially true for the filterbank features. A more elaborated visualization method, based on a top-down generative process in the reverse direction of the classification network, was recently developed by Zeiler and Fergus [436] for examining what features the deep convolutional networks capture from the image data. The power of the deep networks is shown to be their ability to extract appropriate features and do discrimination jointly [210].

As another way to concisely introduce the DNN, we can review the history of artificial neural networks using a "hype cycle," which is a graphic representation of the maturity, adoption and social application of specific technologies. The 2012 version of the hype cycles graph compiled by Gartner is shown in Figure 2.1. It intends to show how a technology or application will evolve over time (according to five phases: technology trigger, peak of inflated expectations, trough of disillusionment, slope of enlightenment, and plateau of production), and to provide a source of insight to manage its deployment.

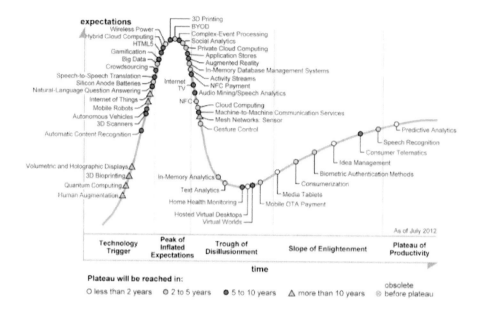

Figure 2.1: Gartner hyper cycle graph representing five phases of a technology (http://en.wikipedia.org/wiki/Hype_cycle).

Applying the Gartner hyper cycle to the artificial neural network development, we created Figure 2.2 to align different generations of the neural network with the various phases designated in the hype cycle. The peak activities ("expectations" or "media hype" on the vertical axis) occurred in late 1980s and early 1990s, corresponding to the height of what is often referred to as the "second generation" of neural networks. The deep belief network (DBN) and a fast algorithm for training it were invented in 2006 [163, 164]. When the DBN was used to initialize the DNN, the learning became highly effective and this has inspired the subsequent fast growing research ("enlightenment" phase shown in Figure 2.2). Applications of the DBN and DNN to industry-scale speech feature extraction and speech recognition started in 2009 when leading academic and industrial researchers with both deep learning and speech expertise collaborated; see reviews in [89, 161]. This collaboration fast expanded the work of speech recognition using deep learning methods to increasingly larger successes [94, 161, 323, 414],

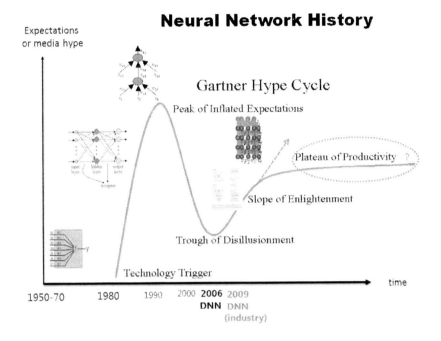

Figure 2.2: Applying Gartner hyper cycle graph to analyzing the history of artificial neural network technology (We thank our colleague John Platt during 2012 for bringing this type of "Hyper Cycle" graph to our attention for concisely analyzing the neural network history).

many of which will be covered in the remainder of this monograph. The height of the "plateau of productivity" phase, not yet reached in our opinion, is expected to be higher than that in the stereotypical curve (circled with a question mark in Figure 2.2), and is marked by the dashed line that moves straight up.

We show in Figure 2.3 the history of speech recognition, which has been compiled by NIST, organized by plotting the word error rate (WER) as a function of time for a number of increasingly difficult speech recognition tasks. Note all WER results were obtained using the GMM–HMM technology. When one particularly difficult task (Switchboard) is extracted from Figure 2.3, we see a flat curve over many years using the GMM–HMM technology but after the DNN technology is used the WER drops sharply (marked by the red star in Figure 2.4).

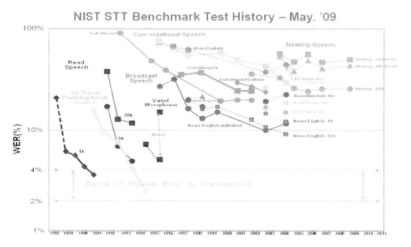

Figure 2.3: The famous NIST plot showing the historical speech recognition error rates achieved by the GMM-HMM approach for a number of increasingly difficult speech recognition tasks. Data source: http://itl.nist.gov/iad/mig/publications/ASRhistory/index.html

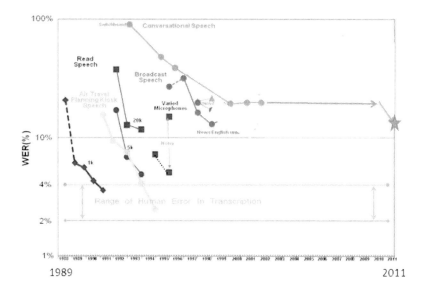

Figure 2.4: Extracting WERs of one task from Figure 2.3 and adding the significantly lower WER (marked by the star) achieved by the DNN technology.

In the next section, an overview is provided on the various architectures of deep learning, followed by more detailed expositions of a few widely studied architectures and methods and by selected applications in signal and information processing including speech and audio, natural language, information retrieval, vision, and multi-modal processing.

3

Three Classes of Deep Learning Networks

3.1 A three-way categorization

As described earlier, deep learning refers to a rather wide class of machine learning techniques and architectures, with the hallmark of using many layers of non-linear information processing that are hierarchical in nature. Depending on how the architectures and techniques are intended for use, e.g., synthesis/generation or recognition/ classification, one can broadly categorize most of the work in this area into three major classes:

1. **Deep networks for unsupervised or generative learning**, which are intended to capture high-order correlation of the observed or visible data for pattern analysis or synthesis purposes when no information about target class labels is available. Unsupervised feature or representation learning in the literature refers to this category of the deep networks. When used in the generative mode, may also be intended to characterize joint statistical distributions of the visible data and their associated classes when available and being treated as part of the visible data. In the

latter case, the use of Bayes rule can turn this type of generative networks into a discriminative one for learning.

2. **Deep networks for supervised learning**, which are intended to directly provide discriminative power for pattern classification purposes, often by characterizing the posterior distributions of classes conditioned on the visible data. Target label data are always available in direct or indirect forms for such supervised learning. They are also called discriminative deep networks.

3. **Hybrid deep networks**, where the goal is discrimination which is assisted, often in a significant way, with the outcomes of generative or unsupervised deep networks. This can be accomplished by better optimization or/and regularization of the deep networks in category (2). The goal can also be accomplished when discriminative criteria for supervised learning are used to estimate the parameters in any of the deep generative or unsupervised deep networks in category (1) above.

Note the use of "hybrid" in (3) above is different from that used sometimes in the literature, which refers to the hybrid systems for speech recognition feeding the output probabilities of a neural network into an HMM [17, 25, 42, 261].

By the commonly adopted machine learning tradition (e.g., Chapter 28 in [264], and Reference [95], it may be natural to just classify deep learning techniques into deep discriminative models (e.g., deep neural networks or DNNs, recurrent neural networks or RNNs, convolutional neural networks or CNNs, etc.) and generative/unsupervised models (e.g., restricted Boltzmann machine or RBMs, deep belief networks or DBNs, deep Boltzmann machines (DBMs), regularized autoencoders, etc.). This two-way classification scheme, however, misses a key insight gained in deep learning research about how generative or unsupervised-learning models can greatly improve the training of DNNs and other deep discriminative or supervised-learning models via better regularization or optimization. Also, deep networks for unsupervised learning may not necessarily need to be probabilistic or be able to meaningfully sample from the model (e.g., traditional autoencoders, sparse coding networks, etc.). We note here that more recent studies have generalized the traditional denoising autoencoders so that they can be efficiently sampled from and thus have become generative models [5, 24, 30]. Nevertheless, the traditional two-way classification indeed points to several key differences between deep networks for unsupervised and supervised learning. Compared between the two, deep supervised-learning models such as DNNs are usually more efficient to train and test, more flexible to construct, and more suitable for end-to-end learning of complex systems (e.g., no approximate inference and learning such as loopy belief propagation). On the other hand, the deep unsupervised-learning models, especially the probabilistic generative ones, are easier to interpret, easier to embed domain knowledge, easier to compose, and easier to handle uncertainty, but they are typically intractable in inference and learning for complex systems. These distinctions are retained also in the proposed three-way classification which is hence adopted throughout this monograph.

Below we review representative work in each of the above three categories, where several basic definitions are summarized in Table 3.1. Applications of these deep architectures, with varied ways of learning including supervised, unsupervised, or hybrid, are deferred to Sections 7–11.

3.2 Deep networks for unsupervised or generative learning

Unsupervised learning refers to no use of task specific supervision information (e.g., target class labels) in the learning process. Many deep networks in this category can be used to meaningfully generate samples by sampling from the networks, with examples of RBMs, DBNs, DBMs, and generalized denoising autoencoders [23], and are thus generative models. Some networks in this category, however, cannot be easily sampled, with examples of sparse coding networks and the original forms of deep autoencoders, and are thus not generative in nature.

Among the various subclasses of generative or unsupervised deep networks, the energy-based deep models are the most common [28, 20, 213, 268]. The original form of the deep autoencoder [28, 100, 164], which we will give more detail about in Section 4, is a typical example

Table 3.1: Basic deep learning terminologies.

Deep Learning: a class of machine learning techniques, where many layers of information processing stages in hierarchical supervised architectures are exploited for unsupervised feature learning and for pattern analysis/classification. The essence of deep learning is to compute hierarchical features or representations of the observational data, where the higher-level features or factors are defined from lower-level ones. The family of deep learning methods have been growing increasingly richer, encompassing those of neural networks, hierarchical probabilistic models, and a variety of unsupervised and supervised feature learning algorithms.

Deep belief network (DBN): probabilistic generative models composed of multiple layers of stochastic, hidden variables. The top two layers have undirected, symmetric connections between them. The lower layers receive top-down, directed connections from the layer above.

Boltzmann machine (BM): a network of symmetrically connected, neuron-like units that make stochastic decisions about whether to be on or off.

Restricted Boltzmann machine (RBM): a special type of BM consisting of a layer of visible units and a layer of hidden units with no visible-visible or hidden-hidden connections.

Deep neural network (DNN): a multilayer perceptron with many hidden layers, whose weights are fully connected and are often (although not always) initialized using either an unsupervised or a supervised pretraining technique. (In the literature prior to 2012, a DBN was often used incorrectly to mean a DNN.)

Deep autoencoder: a "discriminative" DNN whose output targets are the data input itself rather than class labels; hence an unsupervised learning model. When trained with a denoising criterion, a deep autoencoder is also a generative model and can be sampled from.

(Continued)

Table 3.1: (*Continued*)

Distributed representation: an internal representation of the observed data in such a way that they are modeled as being explained by the interactions of many hidden factors. A particular factor learned from configurations of other factors can often generalize well to new configurations. Distributed representations naturally occur in a "connectionist" neural network, where a concept is represented by a pattern of activity across a number of units and where at the same time a unit typically contributes to many concepts. One key advantage of such many-to-many correspondence is that they provide robustness in representing the internal structure of the data in terms of graceful degradation and damage resistance. Another key advantage is that they facilitate generalizations of concepts and relations, thus enabling reasoning abilities.

---

of this unsupervised model category. Most other forms of deep autoencoders are also unsupervised in nature, but with quite different properties and implementations. Examples are transforming autoencoders [160], predictive sparse coders and their stacked version, and de-noising autoencoders and their stacked versions [376].

Specifically, in de-noising autoencoders, the input vectors are first corrupted by, for example, randomly selecting a percentage of the inputs and setting them to zeros or adding Gaussian noise to them. Then the parameters are adjusted for the hidden encoding nodes to reconstruct the original, uncorrupted input data using criteria such as mean square reconstruction error and KL divergence between the original inputs and the reconstructed inputs. The encoded representations transformed from the uncorrupted data are used as the inputs to the next level of the stacked de-noising autoencoder.

Another prominent type of deep unsupervised models with generative capability is the deep Boltzmann machine or DBM [131, 315, 316, 348]. A DBM contains many layers of hidden variables, and has no connections between the variables within the same layer. This is a special case of the general Boltzmann machine (BM), which is a network of symmetrically connected units that are on or off based on a stochastic mechanism. While having a simple learning algorithm, the general BMs are very complex to study and very slow to train. In a DBM, each layer captures complicated, higher-order correlations between the activities of hidden features in the layer below. DBMs have the potential of learning internal representations that become increasingly complex, highly desirable for solving object and speech recognition problems. Further, the high-level representations can be built from a large supply of unlabeled sensory inputs and very limited labeled data can then be used to only slightly fine-tune the model for a specific task at hand.

When the number of hidden layers of DBM is reduced to one, we have restricted Boltzmann machine (RBM). Like DBM, there are no hidden-to-hidden and no visible-to-visible connections in the RBM. The main virtue of RBM is that via composing many RBMs, many hidden layers can be learned efficiently using the feature activations of one RBM as the training data for the next. Such composition leads to deep belief network (DBN), which we will describe in more detail, together with RBMs, in Section 5.

The standard DBN has been extended to the factored higher-order Boltzmann machine in its bottom layer, with strong results obtained for phone recognition [64] and for computer vision [296]. This model, called the mean-covariance RBM or mcRBM, recognizes the limitation of the standard RBM in its ability to represent the covariance structure of the data. However, it is difficult to train mcRBMs and to use them at the higher levels of the deep architecture. Further, the strong results published are not easy to reproduce. In the architecture described by Dahl et al. [64], the mcRBM parameters in the full DBN are not fine-tuned using the discriminative information, which is used for fine tuning the higher layers of RBMs, due to the high computational cost. Subsequent work showed that when speaker adapted features are used, which remove more variability in the features, mcRBM was not helpful [259].

Another representative deep generative network that can be used for unsupervised (as well as supervised) learning is the sum–product network or SPN [125, 289]. An SPN is a directed acyclic graph with the observed variables as leaves, and with sum and product operations as internal nodes in the deep network. The "sum" nodes give mixture models, and the "product" nodes build up the feature hierarchy. Properties of "completeness" and "consistency" constrain the SPN in a desirable way. The learning of SPNs is carried out using the EM algorithm together with back-propagation. The learning procedure starts with a dense SPN. It then finds an SPN structure by learning its weights, where zero weights indicate removed connections. The main difficulty in learning SPNs is that the learning signal (i.e., the gradient) quickly dilutes when it propagates to deep layers. Empirical solutions have been found to mitigate this difficulty as reported in [289]. It was pointed out in that early paper that despite the many desirable generative properties in the SPN, it is difficult to fine tune the parameters using the discriminative information, limiting its effectiveness in classification tasks. However, this difficulty has been overcome in the subsequent work reported in [125], where an efficient BP-style discriminative training algorithm for SPN was presented. Importantly, the standard gradient descent, based on the derivative of the conditional likelihood, suffers from the same gradient diffusion problem well known in the regular DNNs. The trick to alleviate this problem in learning SPNs is to replace the marginal inference with the most probable state of the hidden variables and to propagate gradients through this "hard" alignment only. Excellent results on small-scale image recognition tasks were reported by Gens and Domingo [125].

Recurrent neural networks (RNNs) can be considered as another class of deep networks for unsupervised (as well as supervised) learning, where the depth can be as large as the length of the input data sequence. In the unsupervised learning mode, the RNN is used to predict the data sequence in the future using the previous data samples, and no additional class information is used for learning. The RNN is very powerful for modeling sequence data (e.g., speech or text), but until recently they had not been widely used partly because they are difficult to train to capture long-term dependencies, giving rise to gradient vanishing or gradient explosion problems which were known in early 1990s [29, 167]. These problems can now be dealt with more easily [24, 48, 85, 280]. Recent advances in Hessian-free optimization [238] have also partially overcome this difficulty using approximated second-order information or stochastic curvature estimates. In the more recent work [239], RNNs that are trained with Hessian-free optimization are used as a generative deep network in the character-level language modeling tasks, where gated connections are introduced to allow the current input characters to predict the transition from one latent state vector to the next. Such generative RNN models are demonstrated to be well capable of generating sequential text characters. More recently, Bengio et al. [22] and Sutskever [356] have explored variations of stochastic gradient descent optimization algorithms in training generative RNNs and shown that these algorithms can outperform Hessian-free optimization methods. Mikolov et al. [248] have reported excellent results on using RNNs for language modeling. Most recently, Mesnil et al. [242] and Yao et al. [403] reported the success of RNNs in spoken language understanding. We will review this set of work in Section 8.

There has been a long history in speech recognition research where human speech production mechanisms are exploited to construct dynamic and deep structure in probabilistic generative models; for a comprehensive review, see the monograph by Deng [76]. Specifically, the early work described in [71, 72, 83, 84, 99, 274] generalized and extended the conventional shallow and conditionally independent HMM structure by imposing dynamic constraints, in the form of polynomial trajectory, on the HMM parameters. A variant of this approach has been more recently developed using different learning techniques for time-varying HMM parameters and with the applications extended to speech recognition robustness [431, 416]. Similar trajectory HMMs also form the basis for parametric speech synthesis [228, 326, 439, 438]. Subsequent work added a new hidden layer into the dynamic model to explicitly account for the target-directed, articulatory-like properties in human speech generation [45, 73, 74, 83, 96, 75, 90, 231, 232, 233, 251, 282]. More efficient implementation of this deep architecture with hidden dynamics is achieved with non-recursive or finite impulse response (FIR) filters in more recent studies [76, 107, 105]. The above deep-structured generative models of speech can be shown as special cases of the more general dynamic network model and even more general dynamic graphical models [35, 34]. The graphical models can comprise many hidden layers to characterize the complex relationship between the variables in speech generation. Armed with powerful graphical modeling tool, the deep architecture of speech has more recently been successfully applied to solve the very difficult problem of single-channel, multi-talker speech recognition, where the mixed speech is the visible variable while the un-mixed speech becomes represented in a new hidden layer in the deep generative architecture [301, 391]. Deep generative graphical models are indeed a powerful tool in many applications due to their capability of embedding domain knowledge. However, they are often used with inappropriate approximations in inference, learning, prediction, and topology design, all arising from inherent intractability in these tasks for most real-world applications. This problem has been addressed in the recent work of Stoyanov et al. [352], which provides an interesting direction for making deep generative graphical models potentially more useful in practice in the future. An even more drastic way to deal with this intractability was proposed recently by Bengio et al. [30], where the need to marginalize latent variables is avoided altogether.

The standard statistical methods used for large-scale speech recognition and understanding combine (shallow) hidden Markov models for speech acoustics with higher layers of structure representing different levels of natural language hierarchy. This combined hierarchical model can be suitably regarded as a deep generative architecture, whose motivation and some technical detail may be found in Section 7 of the recent monograph [200] on "Hierarchical HMM" or HHMM. Related models with greater technical depth and mathematical treatment can be found in [116] for HHMM and [271] for Layered HMM. These early deep models were formulated as directed graphical models, missing the key aspect of "distributed representation" embodied in the more recent deep generative networks of the DBN and DBM discussed earlier in this chapter. Filling in this missing aspect would help improve these generative models.

Finally, dynamic or temporally recursive generative models based on neural network architectures can be found in [361] for human motion modeling, and in [344, 339] for natural language and natural scene parsing. The latter model is particularly interesting because the learning algorithms are capable of automatically determining the optimal model structure. This contrasts with other deep architectures such as DBN where only the parameters are learned while the architectures need to be pre-defined. Specifically, as reported in [344], the recursive structure commonly found in natural scene images and in natural language sentences can be discovered using a max-margin structure prediction architecture. It is shown that the units contained in the images or sentences are identified, and the way in which these units interact with each other to form the whole is also identified.

3.3   Deep networks for supervised learning

Many of the discriminative techniques for supervised learning in signal and information processing are shallow architectures such as HMMs [52, 127, 147, 186, 188, 290, 394, 418] and conditional random fields (CRFs) [151, 155, 281, 400, 429, 446]. A CRF is intrinsically a shallow discriminative architecture, characterized by the linear relationship between the input features and the transition features. The shallow nature of the CRF is made most clear by the equivalence established between the CRF and the discriminatively trained Gaussian models and HMMs [148]. More recently, deep-structured CRFs have been developed by stacking the output in each lower layer of the CRF, together with the original input data, onto its higher layer [428]. Various versions of deep-structured CRFs are successfully applied to phone recognition [410], spoken language identification [428], and natural language processing [428]. However, at least for the phone recognition task, the performance of deep-structured CRFs, which are purely discriminative (non-generative), has not been able to match that of the hybrid approach involving DBN, which we will take on shortly.

Morgan [261] gives an excellent review on other major existing discriminative models in speech recognition based mainly on the traditional neural network or MLP architecture using back-propagation learning with random initialization. It argues for the importance of both the increased width of each layer of the neural networks and the increased depth. In particular, a class of deep neural network models forms the basis of the popular "tandem" approach [262], where the output of the discriminatively learned neural network is treated as part of the observation variable in HMMs. For some representative recent work in this area, see [193, 283].

In more recent work of [106, 110, 218, 366, 377], a new deep learning architecture, sometimes called deep stacking network (DSN), together with its tensor variant [180, 181] and its kernel version [102], are developed that all focus on discrimination with scalable, parallelizable, block-wise learning relying on little or no generative component. We will describe this type of discriminative deep architecture in detail in Section 6.

As discussed in the preceding section, recurrent neural networks (RNNs) have been used as a generative model; see also the neural predictive model [87] with a similar "generative" mechanism. RNNs can also be used as a discriminative model where the output is a label sequence associated with the input data sequence. Note that such discriminative RNNs or sequence models were applied to speech a long time ago with limited success. In [17], an HMM was trained jointly with the neural networks, with a discriminative probabilistic training criterion. In [304], a separate HMM was used to segment the sequence during training, and the HMM was also used to transform the RNN classification results into label sequences. However, the use of the HMM for these purposes does not take advantage of the full potential of RNNs.

A set of new models and methods were proposed more recently in [133, 134, 135, 136] that enable the RNNs themselves to perform sequence classification while embedding the long-short-term memory into the model, removing the need for pre-segmenting the training data and for post-processing the outputs. Underlying this method is the idea of interpreting RNN outputs as the conditional distributions over all possible label sequences given the input sequences. Then, a differentiable objective function can be derived to optimize these conditional distributions over the correct label sequences, where the segmentation of the data is performed automatically by the algorithm. The effectiveness of this method has been demonstrated in handwriting recognition tasks and in a small speech task [135, 136] to be discussed in more detail in Section 7 of this monograph.

Another type of discriminative deep architecture is the convolutional neural network (CNN), in which each module consists of a convolutional layer and a pooling layer. These modules are often stacked up with one on top of another, or with a DNN on top of it, to form a deep model [212]. The convolutional layer shares many weights, and the pooling layer subsamples the output of the convolutional layer and reduces the data rate from the layer below. The weight sharing in the convolutional layer, together with appropriately chosen pooling schemes, endows the CNN with some "invariance" properties (e.g., translation invariance). It has been argued that such limited "invariance" or equi-variance is not adequate for complex pattern recognition tasks and more principled ways of handling a wider range of invariance may be needed [160]. Nevertheless, CNNs have been found highly effective and been commonly used in computer vision and image recognition [54, 55, 56, 57, 69, 198, 209, 212, 434]. More recently, with appropriate changes from the CNN designed for image analysis to that taking into account speech-specific properties, the CNN is also found effective for speech recognition [1, 2, 3, 81, 94, 312]. We will discuss such applications in more detail in Section 7 of this monograph.

It is useful to point out that the time-delay neural network (TDNN) [202, 382] developed for early speech recognition is a special case and predecessor of the CNN when weight sharing is limited to one of the two dimensions, i.e., time dimension, and there is no pooling layer. It was not until recently that researchers have discovered that the time-dimension invariance is less important than the frequency-dimension invariance for speech recognition [1, 3, 81]. A careful analysis on the underlying reasons is described in [81], together with a new strategy for designing the CNN's pooling layer demonstrated to be more effective than all previous CNNs in phone recognition.

It is also useful to point out that the model of hierarchical temporal memory (HTM) [126, 143, 142] is another variant and extension of the CNN. The extension includes the following aspects: (1) Time or temporal dimension is introduced to serve as the "supervision" information for discrimination (even for static images); (2) Both bottom-up and top-down information flows are used, instead of just bottom-up in the CNN; and (3) A Bayesian probabilistic formalism is used for fusing information and for decision making.

Finally, the learning architecture developed for bottom-up, detection-based speech recognition proposed in [214] and developed further since 2004, notably in [330, 332, 427] using the DBN–DNN technique, can also be categorized in the discriminative or supervised-learning deep architecture category. There is no intent and mechanism in this architecture to characterize the joint probability of data and recognition targets of speech attributes and of the higher-level phone and words. The most current implementation of this approach is based on the DNN, or neural networks with many layers using back-propagation learning. One intermediate neural network layer in the implementation of this detection-based framework explicitly represents the speech attributes, which are simplified entities from the "atomic" units of speech developed in the early work of [101, 355]. The simplification lies in the removal of the temporally overlapping properties of the speech attributes or articulatory-like features. Embedding such more realistic properties in the future work is expected to improve the accuracy of speech recognition further.

3.4 Hybrid deep networks

The term "hybrid" for this third category refers to the deep architecture that either comprises or makes use of both generative and discriminative model components. In the existing hybrid architectures published in the literature, the generative component is mostly exploited to help with discrimination, which is the final goal of the hybrid architecture. How and why generative modeling can help with discrimination can be examined from two viewpoints [114]:

- The optimization viewpoint where generative models trained in an unsupervised fashion can provide excellent initialization points in highly nonlinear parameter estimation problems (The commonly used term of "pre-training" in deep learning has been introduced for this reason); and/or

- The regularization perspective where the unsupervised-learning models can effectively provide a prior on the set of functions representable by the model.

The study reported in [114] provided an insightful analysis and experimental evidence supporting both of the viewpoints above.

The DBN, a generative, deep network for unsupervised learning discussed in Section 3.2, can be converted to and used as the initial model of a DNN for supervised learning with the same network structure, which is further discriminatively trained or fine-tuned using the target labels provided. When the DBN is used in this way we consider this DBN–DNN model as a hybrid deep model, where the model trained using unsupervised data helps to make the discriminative model effective for supervised learning. We will review details of the discriminative DNN for supervised learning in the context of RBM/DBN generative, unsupervised pre-training in Section 5.

Another example of the hybrid deep network is developed in [260], where the DNN weights are also initialized from a generative DBN but are further fine-tuned with a sequence-level discriminative criterion, which is the conditional probability of the label sequence given the input feature sequence, instead of the frame-level criterion of cross-entropy commonly used. This can be viewed as a combination of the static DNN with the shallow discriminative architecture of CRF. It can be shown that such a DNN–CRF is equivalent to a hybrid deep architecture of DNN and HMM whose parameters are learned jointly using the full-sequence maximum mutual information (MMI) criterion between the entire label sequence and the input feature sequence. A closely related full-sequence training method designed and implemented for much larger tasks is carried out more recently with success for a shallow neural network [194] and for a deep one [195, 353, 374]. We note that the origin of the idea for joint training of the sequence model (e.g., the HMM) and of the neural network came from the early work of [17, 25], where shallow neural networks were trained with small amounts of training data and with no generative pre-training.

Here, it is useful to point out a connection between the above pretraining/fine-tuning strategy associated with hybrid deep networks and the highly popular minimum phone error (MPE) training technique for the HMM (see [147, 290] for an overview). To make MPE training effective, the parameters need to be initialized using an algorithm (e.g., Baum-Welch algorithm) that optimizes a generative criterion (e.g., maximum likelihood). This type of methods, which uses maximum-likelihood trained parameters to assist in the discriminative HMM training can be viewed as a "hybrid" approach to train the shallow HMM model.

Along the line of using discriminative criteria to train parameters in generative models as in the above HMM training example, we here discuss the same method applied to learning other hybrid deep networks. In [203], the generative model of RBM is learned using the discriminative criterion of posterior class-label probabilities. Here the label vector is concatenated with the input data vector to form the combined visible layer in the RBM. In this way, RBM can serve as a stand-alone solution to classification problems and the authors derived a discriminative learning algorithm for RBM as a shallow generative model. In the more recent work by Ranzato et al. [298], the deep generative model of DBN with gated Markov random field (MRF) at the lowest level is learned for feature extraction and then for recognition of difficult image classes including occlusions. The generative ability of the DBN facilitates the discovery of what information is captured and what is lost at each level of representation in the deep model, as demonstrated in [298]. A related study on using the discriminative criterion of empirical risk to train deep graphical models can be found in [352].

A further example of hybrid deep networks is the use of generative models of DBNs to pre-train deep convolutional neural networks (deep CNNs) [215, 216, 217]. Like the fully connected DNN discussed earlier, pre-training also helps to improve the performance of deep CNNs over random initialization. Pre-training DNNs or CNNs using a set of regularized deep autoencoders [24], including denoising autoencoders, contractive autoencoders, and sparse autoencoders, is also a similar example of the category of hybrid deep networks.

The final example given here for hybrid deep networks is based on the idea and work of [144, 267], where one task of discrimination (e.g., speech recognition) produces the output (text) that serves as the input to the second task of discrimination (e.g., machine translation). The overall system, giving the functionality of speech translation — translating speech in one language into text in another language — is a two-stage deep architecture consisting of both generative and discriminative elements. Both models of speech recognition (e.g., HMM) and of machine translation (e.g., phrasal mapping and non-monotonic alignment) are generative in nature, but their parameters are all learned for discrimination of the ultimate translated text given the speech data. The framework described in [144] enables end-to-end performance optimization in the overall deep architecture using the unified learning framework initially published in [147]. This hybrid deep learning approach can be applied to not only speech translation but also all speech-centric and possibly other information processing tasks such as speech information retrieval, speech understanding, cross-lingual speech/text understanding and retrieval, etc. (e.g., [88, 94, 145, 146, 366, 398]).

In the next three chapters, we will elaborate on three prominent types of models for deep learning, one from each of the three classes reviewed in this chapter. These are chosen to serve the tutorial purpose, given their simplicity of the architectural and mathematical descriptions. The three architectures described in the following three chapters may not be interpreted as the most representative and influential work in each of the three classes.

4

Deep Autoencoders — Unsupervised Learning

This section and the next two will each select one prominent example deep network for each of the three categories outlined in Section 3. Here we begin with the category of the deep models designed mainly for unsupervised learning.

4.1 Introduction

The deep autoencoder is a special type of the DNN (with no class labels), whose output vectors have the same dimensionality as the input vectors. It is often used for learning a representation or effective encoding of the original data, in the form of input vectors, at hidden layers. Note that the autoencoder is a nonlinear feature extraction method without using class labels. As such, the features extracted aim at conserving and better representing information instead of performing classification tasks, although sometimes these two goals are correlated.

An autoencoder typically has an input layer which represents the original data or input feature vectors (e.g., pixels in image or spectra in speech), one or more hidden layers that represent the transformed feature, and an output layer which matches the input layer for reconstruction. When the number of hidden layers is greater than one, the autoencoder is considered to be deep. The dimension of the hidden layers can be either smaller (when the goal is feature compression) or larger (when the goal is mapping the feature to a higher-dimensional space) than the input dimension.

An autoencoder is often trained using one of the many back-propagation variants, typically the stochastic gradient descent method. Though often reasonably effective, there are fundamental problems when using back-propagation to train networks with many hidden layers. Once the errors get back-propagated to the first few layers, they become minuscule, and training becomes quite ineffective. Though more advanced back-propagation methods help with this problem to some degree, it still results in slow learning and poor solutions, especially with limited amounts of training data. As mentioned in the previous chapters, the problem can be alleviated by pre-training each layer as a simple autoencoder [28, 163]. This strategy has been applied to construct a deep autoencoder to map images to short binary code for fast, content-based image retrieval, to encode documents (called semantic hashing), and to encode spectrogram-like speech features which we review below.

4.2   Use of deep autoencoders to extract speech features

Here we review a set of work, some of which was published in [100], in developing an autoencoder for extracting binary speech codes from the raw speech spectrogram data in an unsupervised manner (i.e., no speech class labels). The discrete representations in terms of a binary code extracted by this model can be used in speech information retrieval or as bottleneck features for speech recognition.

A deep generative model of patches of spectrograms that contain 256 frequency bins and 1, 3, 9, or 13 frames is illustrated in Figure 4.1. An undirected graphical model called a Gaussian-Bernoulli RBM is built that has one visible layer of linear variables with Gaussian noise and one hidden layer of 500 to 3000 binary latent variables. After learning the Gaussian-Bernoulli RBM, the activation

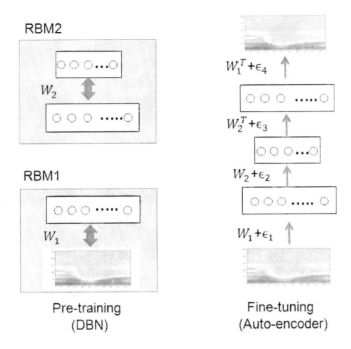

Pre-training
(DBN)

Fine-tuning
(Auto-encoder)

Figure 4.1: The architecture of the deep autoencoder used in [100] for extracting binary speech codes from high-resolution spectrograms. [after [100], @Elsevier].

probabilities of its hidden units are treated as the data for training another Bernoulli-Bernoulli RBM. These two RBM's can then be composed to form a deep belief net (DBN) in which it is easy to infer the states of the second layer of binary hidden units from the input in a single forward pass. The DBN used in this work is illustrated on the left side of Figure 4.1, where the two RBMs are shown in separate boxes. (See more detailed discussions on the RBM and DBN in Section 5).

The deep autoencoder with three hidden layers is formed by "unrolling" the DBN using its weight matrices. The lower layers of this deep autoencoder use the matrices to encode the input and the upper layers use the matrices in reverse order to decode the input. This deep autoencoder is then fine-tuned using error back-propagation to minimize the reconstruction error, as shown on the right side of Figure 4.1. After learning is complete, any variable-length spectrogram can be encoded and reconstructed as follows. First, N consecutive overlapping frames of 256-point log power spectra are each normalized to zero-mean and unit-variance across samples per feature to provide the input to the deep autoencoder. The first hidden layer then uses the logistic function to compute real-valued activations. These real values are fed to the next, coding layer to compute "codes." The real-valued activations of hidden units in the coding layer are quantized to be either zero or one with 0.5 as the threshold. These binary codes are then used to reconstruct the original spectrogram, where individual fixed-frame patches are reconstructed first using the two upper layers of network weights. Finally, the standard overlap-and-add technique in signal processing is used to reconstruct the full-length speech spectrogram from the outputs produced by applying the deep autoencoder to every possible window of N consecutive frames. We show some illustrative encoding and reconstruction examples below.

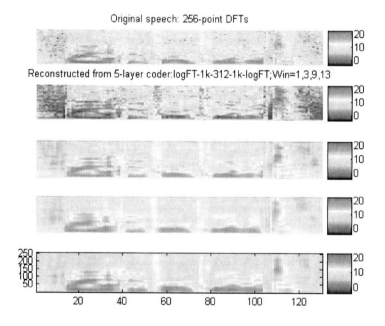

Figure 4.2: Top to Bottom: The ordinal spectrogram; reconstructions using input window sized of $N = 1, 3, 9$, and 13 while forcing the coding units to take values of zero one (i.e., a binary code) . [after [100], @Elsevier].

At the top of Figure 4.2 is the original, un-coded speech, followed by the speech utterances reconstructed from the binary codes (zero or one) at the 312 unit bottleneck code layer with encoding window lengths of $N = 1, 3, 9$, and 13, respectively. The lower reconstruction errors for $N = 9$ and $N = 13$ are clearly seen.

Encoding error of the deep autoencoder is qualitatively examined in comparison with the more traditional codes via vector quantization (VQ). Figure 4.3 shows various aspects of the encoding errors. At the top is the original speech utterance's spectrogram. The next two spectrograms are the blurry reconstruction from the 312-bit VQ and the much more faithful reconstruction from the 312-bit deep autoencoder. Coding errors from both coders, plotted as a function of time, are

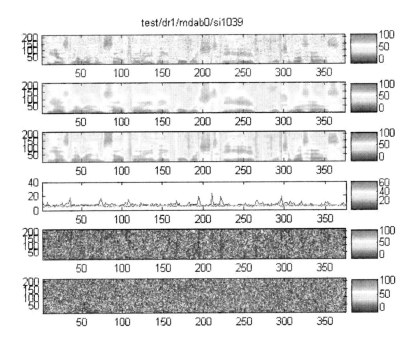

Figure 4.3: Top to bottom: The original spectrogram from the test set; reconstruction from the 312-bit VQ coder; reconstruction from the 312-bit autoencoder; coding errors as a function of time for the VQ coder (blue) and autoencoder (red); spectrogram of the VQ coder residual; spectrogram of the deep autoencoder's residual. [after [100], @ Elsevier].

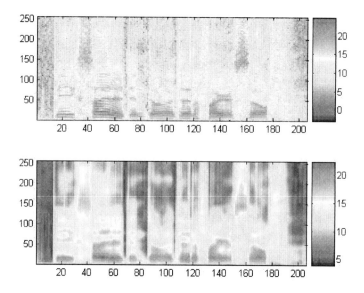

Figure 4.4: The original speech spectrogram and the reconstructed counterpart. A total of 312 binary codes are with one for each single frame.

shown below the spectrograms, demonstrating that the autoencoder (red curve) is producing lower errors than the VQ coder (blue curve) throughout the entire span of the utterance. The final two spectrograms show detailed coding error distributions over both time and frequency bins.

Figures 4.4 to 4.10 show additional examples (unpublished) for the original un-coded speech spectrograms and their reconstructions using the deep autoencoder. They give a diverse number of binary codes for either a single or three consecutive frames in the spectrogram samples.

4.3　Stacked denoising autoencoders

In early years of autoencoder research, the encoding layer had smaller dimensions than the input layer. However, in some applications, it is desirable that the encoding layer is wider than the input layer, in which case techniques are needed to prevent the neural network from learning the trivial identity mapping function. One of the reasons for using a

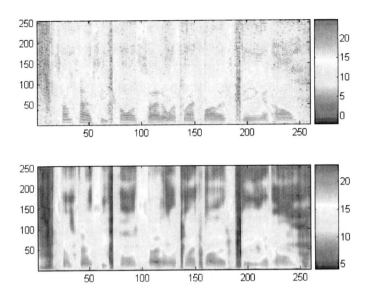

Figure 4.5: Same as Figure 4.4 but with a different TIMIT speech utterance.

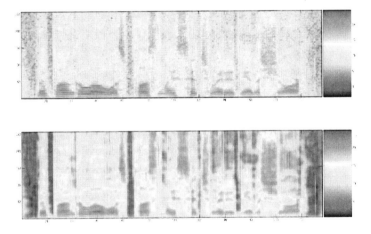

Figure 4.6: The original speech spectrogram and the reconstructed counterpart. A total of 936 binary codes are used for three adjacent frames.

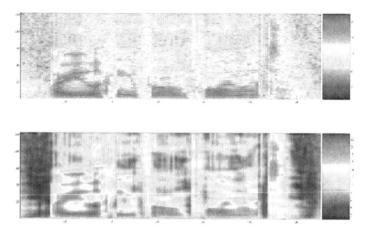

Figure 4.7: Same as Figure 4.6 but with a different TIMIT speech utterance.

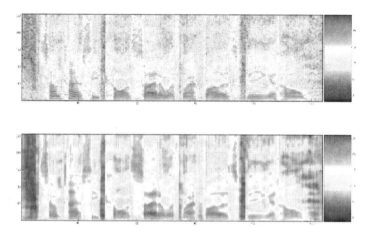

Figure 4.8: Same as Figure 4.6 but with yet another TIMIT speech utterance.

higher dimension in the hidden or encoding layers than the input layer is that it allows the autoencoder to capture a rich input distribution.

The trivial mapping problem discussed above can be prevented by methods such as using sparseness constraints, or using the "dropout" trick by randomly forcing certain values to be zero and thus introducing distortions at the input data [376, 375] or at the hidden layers [166]. For

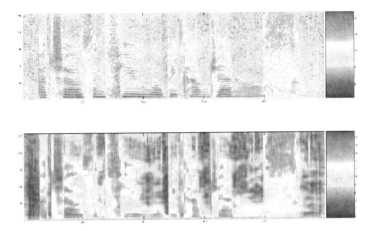

Figure 4.9: The original speech spectrogram and the reconstructed counterpart. A total of 2000 binary codes with one for each single frame.

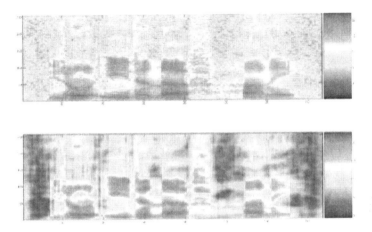

Figure 4.10: Same as Figure 4.9 but with a different TIMIT speech utterance.

example, in the stacked denoising autoencoder detailed in [376], random noises are added to the input data. This serves several purposes. First, by forcing the output to match the original undistorted input data the model can avoid learning the trivial identity solution. Second, since the noises are added randomly, the model learned would be robust to the same kind of distortions in the test data. Third, since each distorted input sample is different, it greatly increases the training set size and thus can alleviate the overfitting problem.

It is interesting to note that when the encoding and decoding weights are forced to be the transpose of each other, such denoising autoencoder with a single sigmoidal hidden layer is strictly equivalent to a particular Gaussian RBM, but instead of training it by the technique of contrastive divergence (CD) or persistent CD, it is trained by a score matching principle, where the score is defined as the derivative of the log-density with respect to the input [375]. Furthermore, Alain and Bengio [5] generalized this result to any parameterization of the encoder and decoder with squared reconstruction error and Gaussian corruption noise. They show that as the amount of noise approaches zero, such models estimate the true score of the underlying data generating distribution. Finally, Bengio et al. [30] show that any denoising autoencoder is a consistent estimator of the underlying data generating distribution within some family of distributions. This is true for any parameterization of the autoencoder, for any type of information-destroying corruption process with no constraint on the noise level except being positive, and for any reconstruction loss expressed as a conditional log-likelihood. The consistency of the estimator is achieved by associating the denoising autoencoder with a Markov chain whose stationary distribution is the distribution estimated by the model, and this Markov chain can be used to sample from the denoising autoencoder.

4.4   Transforming autoencoders

The deep autoencoder described above can extract faithful codes for feature vectors due to many layers of nonlinear processing. However, the code extracted in this way is transformation-variant. In other words, the extracted code would change in ways chosen by the learner when the input feature vector is transformed. Sometimes, it is desirable to have the code change predictably to reflect the underlying transformation-invariant property of the perceived content. This is the goal of the transforming autoencoder proposed in [162] for image recognition.

The building block of the transforming autoencoder is a "capsule," which is an independent sub-network that extracts a single parameterized feature representing a single entity, be it visual or audio. A transforming autoencoder receives both an input vector and a target output vector, which is transformed from the input vector through a simple global transformation mechanism; e.g., translation of an image and frequency shift of speech (the latter due to the vocal tract length difference). An explicit representation of the global transformation is assumed known. The coding layer of the transforming autoencoder consists of the outputs of several capsules.

During the training phase, the different capsules learn to extract different entities in order to minimize the error between the final output and the target.

In addition to the deep autoencoder architectures described here, there are many other types of generative architectures in the literature, all characterized by the use of data alone (i.e., free of classification labels) to automatically derive higher-level features.

5

Pre-Trained Deep Neural Networks — A Hybrid

In this section, we present the most widely used hybrid deep architecture — the pre-trained deep neural network (DNN), and discuss the related techniques and building blocks including the RBM and DBN. We discuss the DNN example here in the category of hybrid deep networks before the examples in the category of deep networks for supervised learning (Section 6). This is partly due to the natural flow from the unsupervised learning models to the DNN as a hybrid model. The discriminative nature of artificial neural networks for supervised learning has been widely known, and thus would not be required for understanding the hybrid nature of the DNN that uses unsupervised pre-training to facilitate the subsequent discriminative fine tuning.

Part of the review in this chapter is based on recent publications in [68, 161, 412].

5.1 Restricted Boltzmann machines

An RBM is a special type of Markov random field that has one layer of (typically Bernoulli) stochastic hidden units and one layer of (typically Bernoulli or Gaussian) stochastic visible or observable units. RBMs can

51

be represented as bipartite graphs, where all visible units are connected to all hidden units, and there are no visible–visible or hidden–hidden connections.

In an RBM, the joint distribution $p(\mathbf{v}, \mathbf{h}; \theta)$ over the visible units $\mathbf{v}$ and hidden units $\mathbf{h}$, given the model parameters θ, is defined in terms of an energy function $E(\mathbf{v}, \mathbf{h}; \theta)$ of

$$p(\mathbf{v}, \mathbf{h}; \theta) = \frac{\exp(-E(\mathbf{v}, \mathbf{h}; \theta))}{Z},$$

where $Z = \sum_{\mathbf{v}} \sum_{\mathbf{h}} \exp(-E(\mathbf{v}, \mathbf{h}; \theta))$ is a normalization factor or partition function, and the marginal probability that the model assigns to a visible vector $\mathbf{v}$ is

$$p(\mathbf{v}; \theta) = \frac{\sum_{\mathbf{h}} \exp(-E(\mathbf{v}, \mathbf{h}; \theta))}{Z}$$

For a Bernoulli (visible)-Bernoulli (hidden) RBM, the energy function is defined as

$$E(\mathbf{v}, \mathbf{h}; \theta) = -\sum_{i=1}^{I} \sum_{j=1}^{J} w_{ij} v_i h_j - \sum_{i=1}^{I} b_i v_i - \sum_{j=1}^{J} a_j h_j.$$

where w_{ij} represents the symmetric interaction term between visible unit v_i and hidden unit h_j, b_i and a_j the bias terms, and I and J are the numbers of visible and hidden units. The conditional probabilities can be efficiently calculated as

$$p(h_j = 1 | \mathbf{v}; \theta) = \sigma\left(\sum_{i=1}^{I} w_{ij} v_i + a_j\right),$$

$$p(v_i = 1 | \mathbf{h}; \theta) = \sigma\left(\sum_{j=1}^{J} w_{ij} h_j + b_i\right),$$

where $\sigma(x) = 1/(1 + \exp(-x))$.

Similarly, for a Gaussian (visible)-Bernoulli (hidden) RBM, the energy is

$$E(\mathbf{v}, \mathbf{h}; \theta) = -\sum_{i=1}^{I} \sum_{j=1}^{J} w_{ij} v_i h_j - \frac{1}{2} \sum_{i=1}^{I} (v_i - b_i)^2 - \sum_{j=1}^{J} a_j h_j,$$

The corresponding conditional probabilities become

$$p(h_j = 1|\mathbf{v};\theta) = \sigma\left(\sum_{i=1}^{I} w_{ij}v_i + a_j\right),$$

$$p(v_i|\mathbf{h};\theta) = N\left(\sum_{j=1}^{J} w_{ij}h_j + b_i, 1\right),$$

where v_i takes real values and follows a Gaussian distribution with mean $\sum_{j=1}^{J} w_{ij}h_j + b_i$ and variance one. Gaussian-Bernoulli RBMs can be used to convert real-valued stochastic variables to binary stochastic variables, which can then be further processed using the Bernoulli-Bernoulli RBMs.

The above discussion used two of the most common conditional distributions for the visible data in the RBM — Gaussian (for continuous-valued data) and binomial (for binary data). More general types of distributions in the RBM can also be used. See [386] for the use of general exponential-family distributions for this purpose.

Taking the gradient of the log likelihood $\log p(\mathbf{v};\theta)$ we can derive the update rule for the RBM weights as:

$$\Delta w_{ij} = E_{\text{data}}(v_i h_j) - E_{\text{model}}(v_i h_j),$$

where $E_{\text{data}}(v_i h_j)$ is the expectation observed in the training set (with h_j sampled given v_i according to the model), and $E_{\text{model}}(v_i h_j)$ is that same expectation under the distribution defined by the model. Unfortunately, $E_{\text{model}}(v_i h_j)$ is intractable to compute. The contrastive divergence (CD) approximation to the gradient was the first efficient method proposed to approximate this expected value, where $E_{\text{model}}(v_i h_j)$ is replaced by running the Gibbs sampler initialized at the data for one or more steps. The steps in approximating $E_{\text{model}}(v_i h_j)$ is summarized as follows:

- Initialize $\mathbf{v_0}$ at data
- Sample $\mathbf{h_0} \sim \mathbf{p(h|v_0)}$
- Sample $\mathbf{v_1} \sim \mathbf{p(v|h_0)}$
- Sample $\mathbf{h_1} \sim \mathbf{p(h|v_1)}$

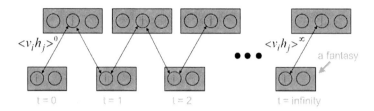

Figure 5.1: A pictorial view of sampling from a RBM during RBM learning (courtesy of Geoff Hinton).

Here, $(\mathbf{v_1}, \mathbf{h_1})$ is a sample from the model, as a very rough estimate of $E_{\text{model}}(v_i h_j)$. The use of $(\mathbf{v_1}, \mathbf{h_1})$ to approximate $E_{\text{model}}(v_i h_j)$ gives rise to the algorithm of CD-1. The sampling process can be pictorially depicted in Figure 5.1.

Note that CD-k generalizes this to more steps of the Markov chain. There are other techniques for estimating the log-likelihood gradient of RBMs, in particular the stochastic maximum likelihood or persistent contrastive divergence (PCD) [363, 406]. Both work better than CD when using the RBM as a generative model.

Careful training of RBMs is essential to the success of applying RBM and related deep learning techniques to solve practical problems. See Technical Report [159] for a very useful practical guide for training RBMs.

The RBM discussed above is both a generative and an unsupervised model, which characterizes the input data distribution using hidden variables and there is no label information involved. However, when the label information is available, it can be used together with the data to form the concatenated "data" set. Then the same CD learning can be applied to optimize the approximate "generative" objective function related to data likelihood. Further, and more interestingly, a "discriminative" objective function can be defined in terms of conditional likelihood of labels. This discriminative RBM can be used to "fine tune" RBM for classification tasks [203].

Ranzato et al. [297, 295] proposed an unsupervised learning algorithm called sparse encoding symmetric machine (SESM), which is quite similar to RBM. They both have a symmetric encoder and decoder, and a logistic nonlinearity on the top of the encoder. The main difference is that whereas the RBM is trained using (very approximate) maximum likelihood, SESM is trained by simply minimizing the average energy plus an additional code sparsity term. SESM relies on the sparsity term to prevent flat energy surfaces, while RBM relies on an explicit contrastive term in the loss, an approximation of the log partition function. Another difference is in the coding strategy in that the code units are "noisy" and binary in the RBM, while they are quasi-binary and sparse in SESM. The use of SESM in pre-training DNNs for speech recognition can be found in [284].

5.2 Unsupervised layer-wise pre-training

Here we describe how to stack up RBMs just described to form a DBN as the basis for DNN's pre-training. Before delving into details, we first note that this procedure, proposed by Hinton and Salakhutdinov [163] is a more general technique of unsupervised layer-wise pretraining. That is, not only RBMs can be stacked to form deep generative (or discriminative) networks, but other types of networks can also do the same, such as autoencoder variants as proposed by Bengio et al. [28].

Stacking a number of the RBMs learned layer by layer from bottom up gives rise to a DBN, an example of which is shown in Figure 5.2. The stacking procedure is as follows. After learning a Gaussian-Bernoulli RBM (for applications with continuous features such as speech) or Bernoulli-Bernoulli RBM (for applications with nominal or binary features such as black-white image or coded text), we treat the activation probabilities of its hidden units as the data for training the Bernoulli-Bernoulli RBM one layer up. The activation probabilities of the second-layer Bernoulli-Bernoulli RBM are then used as the visible data input for the third-layer Bernoulli-Bernoulli RBM, and so on. Some theoretical justification of this efficient layer-by-layer greedy learning strategy is given in [163], where it is shown that the *stacking* procedure above improves a variational lower bound on the likelihood of the training data under the composite model. That is, the greedy procedure

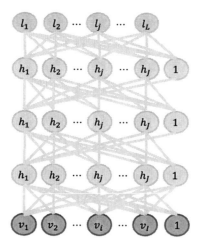

Figure 5.2: An illustration of the DBN-DNN architecture.

above achieves approximate maximum likelihood learning. Note that this learning procedure is unsupervised and requires no class label.

When applied to classification tasks, the generative pre-training can be followed by or combined with other, typically discriminative, learning procedures that fine-tune all of the weights jointly to improve the performance of the network. This discriminative fine-tuning is performed by adding a final layer of variables that represent the desired outputs or labels provided in the training data. Then, the back-propagation algorithm can be used to adjust or fine-tune the network weights in the same way as for the standard feed-forward neural network. What goes to the top, label layer of this DNN depends on the application. For speech recognition applications, the top layer, denoted by "$l_1, l_2, \ldots, l_j, \ldots, l_L$," in Figure 5.2, can represent either syllables, phones, sub-phones, phone states, or other speech units used in the HMM-based speech recognition system.

The generative pre-training described above has produced better phone and speech recognition results than random initialization on a wide variety of tasks, which will be surveyed in Section 7. Further research has also shown the effectiveness of other pre-training strategies. As an example, greedy layer-by-layer training may be carried out with an additional discriminative term to the generative cost function at each level. And without generative pre-training, purely discriminative training of DNNs from random initial weights using the traditional stochastic gradient decent method has been shown to work very well when the scales of the initial weights are set carefully and the mini-batch sizes, which trade off noisy gradients with convergence speed, used in stochastic gradient decent are adapted prudently (e.g., with an increasing size over training epochs). Also, randomization order in creating mini-batches needs to be judiciously determined. Importantly, it was found effective to learn a DNN by starting with a shallow neural network with a single hidden layer. Once this has been trained discriminatively (using early stops to avoid overfitting), a second hidden layer is inserted between the first hidden layer and the labeled softmax output units and the expanded deeper network is again trained discriminatively. This can be continued until the desired number of hidden layers is reached, after which a full backpropagation "fine tuning" is applied. This discriminative "pre-training" procedure is found to work well in practice [324, 419], especially with a reasonably large amount of training data. When the amount of training data is increased even more, then some carefully designed random initialization methods can work well also without using the above pre-training schemes.

In any case, pre-training based on the use of RBMs to stack up in forming the DBN has been found to work well in most cases, regardless of a large or small amount of training data. It is useful to point out that there are other ways to perform pre-training in addition to the use of RBMs and DBNs. For example, denoising autoencoders have now been shown to be consistent estimators of the data generating distribution [30]. Like RBMs, they are also shown to be generative models from which one can sample. Unlike RBMs, however, an unbiased estimator of the gradient of the training objective function can be obtained by the denoising autoencoders, avoiding the need for MCMC or variational approximations in the inner loop of training. Therefore, the greedy layer-wise pre-training may be performed as effectively by stacking the denoising autoencoders as by stacking the RBMs each as a single-layer learner.

Further, a general framework for layer-wise pre-training can be found in many deep learning papers; e.g., Section 2 of [21]. This includes, as a special case, the use of RBMs as the single-layer building block as discussed in this section. The more general framework can cover the RBM/DBN as well as any other unsupervised feature extractor. It can also cover the case of unsupervised pre-training of the representation only followed by a separate stage of learning a classifier on top of the unsupervised, pre-trained features [215, 216, 217].

5.3 Interfacing DNNs with HMMs

The pre-trained DNN as a prominent example of the hybrid deep networks discussed so far in this chapter is a static classifier with input vectors having a fixed dimensionality. However, many practical pattern recognition and information processing problems, including speech recognition, machine translation, natural language understanding, video processing and bio-information processing, require sequence recognition. In sequence recognition, sometimes called classification with structured input/output, the dimensionality of both inputs and outputs are variable.

The HMM, based on dynamic programing operations, is a convenient tool to help port the strength of a static classifier to handle dynamic or sequential patterns. Thus, it is natural to combine feed-forward neural networks and HMMs to bridge the gap between the static and sequence pattern recognition, as was done in the early days of neural networks for speech recognition [17, 25, 42]. A popular architecture to fulfill this role with the use of the DNN is shown in Figure 5.3. This architecture has been successfully used in speech recognition experiments as reported in [67, 68].

It is important to note that the unique elasticity of temporal dynamics of speech as elaborated in [45, 73, 76, 83] would require temporally

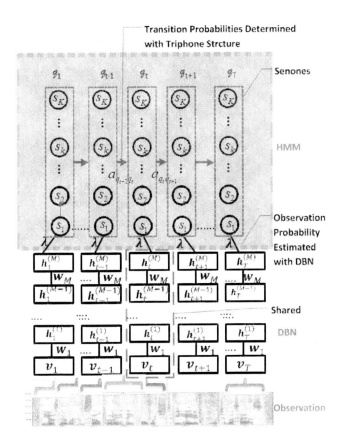

Figure 5.3: Interface between DBN/DNN and HMM to form a DNN–HMM. This architecture, developed at Microsoft, has been successfully used in speech recognition experiments reported in [67, 68]. [after [67, 68], @IEEE].

correlated models more powerful than HMMs for the ultimate success of speech recognition. Integrating such dynamic models that have realistic co-articulatory properties with the DNN and possibly other deep learning models to form the coherent dynamic deep architecture is a challenging new research direction.

6

Deep Stacking Networks and Variants — Supervised Learning

6.1 Introduction

While the DNN just reviewed has been shown to be extremely powerful in connection with performing recognition and classification tasks including speech recognition and image classification, training a DNN has proven to be difficult computationally. In particular, conventional techniques for training DNNs at the fine tuning phase involve the utilization of a stochastic gradient descent learning algorithm, which is difficult to parallelize across machines. This makes learning at large scale nontrivial. For example, it has been possible to use one single, very powerful GPU machine to train DNN-based speech recognizers with dozens to a few hundreds or thousands of hours of speech training data with remarkable results. It is less clear, however, how to scale up this success with much more training data. See [69] for recent work in this direction.

Here we describe a new deep learning architecture, the deep stacking network (DSN), which was originally designed with the learning scalability problem in mind. This chapter is based in part on the recent publications of [106, 110, 180, 181] with expanded discussions.

The central idea of the DSN design relates to the concept of stacking, as proposed and explored in [28, 44, 392], where simple modules of functions or classifiers are composed first and then they are "stacked" on top of each other in order to learn complex functions or classifiers. Various ways of implementing stacking operations have been developed in the past, typically making use of supervised information in the simple modules. The new features for the stacked classifier at a higher level of the stacking architecture often come from concatenation of the classifier output of a lower module and the raw input features. In [60], the simple module used for stacking was a conditional random field (CRF). This type of deep architecture was further developed with hidden states added for successful natural language and speech recognition applications where segmentation information is unknown in the training data [429]. Convolutional neural networks, as in [185], can also be considered as a stacking architecture but the supervision information is typically not used until in the final stacking module.

The DSN architecture was originally presented in [106] and was referred as deep convex network or DCN to emphasize the convex nature of a major portion of the algorithm used for learning the network. The DSN makes use of supervision information for stacking each of the basic modules, which takes the simplified form of multilayer perceptron. In the basic module, the output units are linear and the hidden units are sigmoidal nonlinear. The linearity in the output units permits highly efficient, parallelizable, and closed-form estimation (a result of convex optimization) for the output network weights given the hidden units' activities. Due to the closed-form constraints between the input and output weights, the input weights can also be elegantly estimated in an efficient, parallelizable, batch-mode manner, which we will describe in some detail in Section 6.3.

The name "convex" used in [106] accentuates the role of convex optimization in learning the output network weights given the hidden units' activities in each basic module. It also points to the importance of the closed-form constraints, derived from the convexity, between the input and output weights. Such constraints make the learning of the remaining network parameters (i.e., the input network weights) much easier than otherwise, enabling batch-mode learning of the DSN that can be distributed over CPU clusters. And in more recent publications, the DSN was used when the key operation of stacking is emphasized.

6.2 A basic architecture of the deep stacking network

A DSN, as shown in Figure 6.1, includes a variable number of layered modules, wherein each module is a specialized neural network consisting of a single hidden layer and two trainable sets of weights. In Figure 6.1, only four such modules are illustrated, where each module is shown with a separate color. In practice, up to a few hundreds of modules have been efficiently trained and used in image and speech classification experiments.

The lowest module in the DSN comprises a linear layer with a set of linear input units, a hidden nonlinear layer with a set of nonlinear units, and a second linear layer with a set of linear output units. A sigmoidal nonlinearity is typically used in the hidden layer. However, other nonlinearities can also be used. If the DSN is utilized in connection with recognizing an image, the input units can correspond to a number of pixels (or extracted features) in the image, and can be assigned values based at least in part upon intensity values, RGB values, or the like corresponding to the respective pixels. If the DSN is utilized in connection with speech recognition, the set of input units may correspond to samples of speech waveform, or the extracted features from speech waveforms, such as power spectra or cepstral coefficients. The output units in the linear output layer represent the targets of classification. For instance, if the DSN is configured to perform digit recognition, then the output units may be representative of the values 0, 1, 2, 3, and so forth up to 9 with a 0–1 coding scheme. If the DSN is configured to perform speech recognition, then the output units may be representative of phones, HMM states of phones, or context-dependent HMM states of phones.

The lower-layer weight matrix, which we denote by W, connects the linear input layer and the hidden nonlinear layer. The upper-layer weight matrix, which we denote by U, connects the nonlinear hidden layer with the linear output layer. The weight matrix U can be determined through a closed-form solution given the weight matrix W when the mean square error training criterion is used.

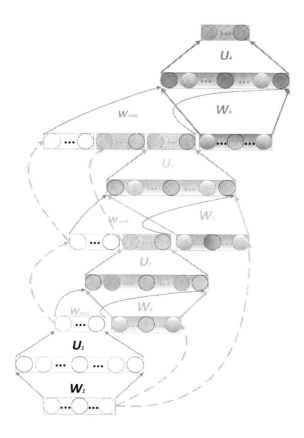

Figure 6.1: A DSN architecture using input–output stacking. Four modules are illustrated, each with a distinct color. Dashed lines denote copying layers. [after [366], @IEEE].

As indicated above, the DSN includes a set of serially connected, overlapping, and layered modules, wherein each module has the same architecture — a linear input layer followed by a nonlinear hidden layer, which is connected to a linear output layer. Note that the output units of a lower module are a subset of the input units of an adjacent higher module in the DSN. More specifically, in a second module that is directly above the lowest module in the DSN, the input units can include the output units of the lowest module and optionally the raw input feature.

This pattern of including output units in a lower module as a portion of the input units in an adjacent higher module and thereafter learning a weight matrix that describes connection weights between hidden units and linear output units via convex optimization can continue for many modules. A resultant learned DSN may then be deployed in connection with an automatic classification task such as frame-level speech phone or state classification. Connecting the DSN's output to an HMM or any dynamic programming device enables continuous speech recognition and other forms of sequential pattern recognition.

6.3 A method for learning the DSN weights

Here, we provide some technical details on how the use of linear output units in the DSN facilitates the learning of the DSN weights. A single module is used to illustrate the advantage for simplicity reasons. First, it is clear that the upper layer weight matrix U can be efficiently learned once the activity matrix H over all training samples in the hidden layer is known. Let's denote the training vectors by $X = [x_1, \ldots, x_i, \ldots, x_N]$, in which each vector is denoted by $x_i = [x_{1i}, \ldots, x_{ji}, \ldots, x_{Di}]^T$ where D is the dimension of the input vector, which is a function of the block, and N is the total number of training samples. Denote by L the number of hidden units and by C the dimension of the output vector. Then the output of a DSN block is $y_i = U^T h_i$ where $h_i = \sigma(W^T x_i)$ is the hidden-layer vector for sample i, U is an $L \times C$ weight matrix at the upper layer of a block. W is a $D \times L$ weight matrix at the lower layer of a block, and $\sigma(\cdot)$ is a sigmoid function. Bias terms are implicitly represented in the above formulation if x_i and h_i are augmented with ones.

Given target vectors in the full training set with a total of N samples, $T = [t_1, \ldots, t_i, \ldots, t_N]$, where each vector is $t_i = [t_{1i}, \cdots, t_{ji}, \ldots, t_{Ci}]^T$, the parameters U and W are learned so as to minimize the average of the total square error below:

$$E = \frac{1}{2} \sum_i \|y_i - t_i\|^2 = \frac{1}{2} \mathrm{Tr}[(Y - T)(Y - T)^T]$$

where the output of the network is

$$y_i = U^T h_i = U^T \sigma(W^T x_i) = G_i(UW)$$

which depends on both weight matrices, as in the standard neural net. Assuming $H = [h_1, \ldots, h_i, \ldots, h_N]$ is known, or equivalently, W is known. Then, setting the error derivative with respective to U to zero gives

$$U = (HH^T)^{-1}HT^T = F(W), \quad \text{where } h_i = \sigma(W^T x_i).$$

This provides an explicit constraint between U and W which were treated independently in the conventional backpropagation algorithm.

Now, given the equality constraint $U = F(W)$, let's use Lagrangian multiplier method to solve the optimization problem in learning W Optimizing the Lagrangian:

$$E = \frac{1}{2}\sum_i \|G_i(U, W) - t_i\|^2 + \lambda\|U - F(W)\|$$

we can derive batch-mode gradient descent learning algorithm where the gradient takes the following form [106, 413]:

$$\frac{\partial E}{\partial W} = 2X[H^T \circ (1 - H)^T \circ [H^\dagger(HT^T)(TH^\dagger) - T^T(TH^\dagger)]],$$

where $H^\dagger = H^T(HH^T)^{-1}$ is pseudo-inverse of H and symbol $\circ$ denotes element-wise multiplication.

Compared with conventional backpropagation, the above method has less noise in gradient computation due to the exploitation of the explicit constraint $U = F(W)$. As such, it was found experimentally that, unlike backpropagation, batch training is effective, which aids parallel learning of the DSN.

6.4 The tensor deep stacking network

The above DSN architecture has recently been generalized to its tensorized version, which we call the tensor DSN (TDSN) [180, 181]. It has the same scalability as the DSN in terms of parallelizability in learning, but it generalizes the DSN by providing higher-order feature interactions missing in the DSN.

The architecture of the TDSN is similar to that of the DSN in the way that stacking operation is carried out. That is, modules of the

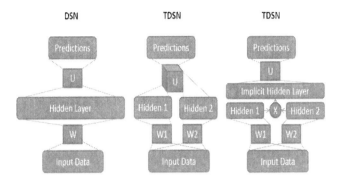

Figure 6.2: Comparisons of a single module of a DSN (left) and that of a tensor DSN (TDSN). Two equivalent forms of a TDSN module are shown to the right. [after [180], @IEEE].

TDSN are stacked up in a similar way to form a deep architecture. The differences between the TDSN and the DSN lie mainly in how each module is constructed. In the DSN, we have one set of hidden units forming a hidden layer, as denoted at the left panel of Figure 6.2. In contrast, each module of a TDSN contains two independent hidden layers, denoted as "Hidden 1" and "Hidden 2" in the middle and right panels of Figure 6.2. As a result of this difference, the upper-layer weights, denoted by "**U**" in Figure 6.2, changes from a matrix (a two dimensional array) in the DSN to a tensor (a three dimensional array) in the TDSN, shown as a cube labeled by "**U**" in the middle panel.

The tensor **U** has a three-way connection, one to the prediction layer and the remaining to the two separate hidden layers. An equivalent form of this TDSN module is shown in the right panel of Figure 6.2, where the implicit hidden layer is formed by expanding the two separate hidden layers into their outer product. The resulting large vector contains all possible pair-wise products for the two sets of hidden-layer vectors. This turns tensor **U** into a matrix again whose dimensions are (1) size of the prediction layer; and (2) product of the two hidden layers' sizes. Such equivalence enables the same convex optimization for learning **U** developed for the DSN to be applied to learning tensor **U**. Importantly, higher-order hidden feature interactions are enabled in the TDSN via the outer product construction for the large, implicit hidden layer.

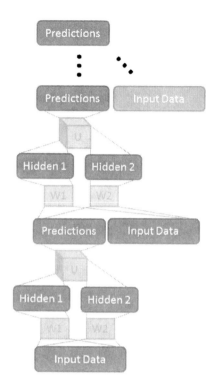

Figure 6.3: Stacking of TDSN modules by concatenating prediction vector with input vector. [after [180], @IEEE].

Stacking the TDSN modules to form a deep architecture pursues in a similar way to the DSN by concatenating various vectors. Two examples are shown in Figures 6.3 and 6.4. Note stacking by concatenating hidden layers with input (Figure 6.4) would be difficult for the DSN since its hidden layer tends to be too large for practical purposes.

6.5 The Kernelized deep stacking network

The DSN architecture has also recently been generalized to its kernelized version, which we call the kernel-DSN (K-DSN) [102, 171]. The motivation of the extension is to increase the size of the hidden units in each DSN module, yet without increasing the size of the free parameters to learn. This goal can be easily accomplished using the kernel trick, resulting in the K-DSN which we describe below.

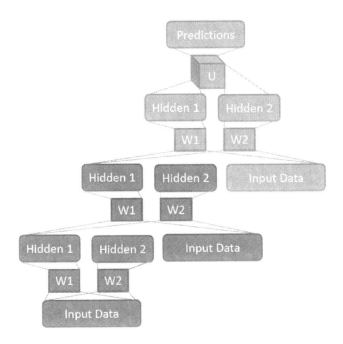

Figure 6.4: Stacking of TDSN modules by concatenating two hidden-layers' vectors with the input vector.

In the DSN architecture reviewed above optimizing the weight matrix U given the hidden layers' outputs in each module is a convex optimization problem. However, the problem of optimizing weight matrix W and thus the whole network is nonconvex. In a recent extension of DSN, a tensor structure was imposed, shifting most of the nonconvex learning burden for W to the convex optimization of U [180, 181]. In the new K-DSN extension, we completely eliminate nonconvex learning for W using the kernel trick.

To derive the K-DSN architecture and the associated learning algorithm, we first take the bottom module of DSN as an example and generalize the sigmoidal hidden layer $\mathbf{h}_i = \sigma(W^{\mathrm{T}} x_i)$ in the DSN module into a generic nonlinear mapping function $G(X)$ from the raw input feature X, with high dimensionality in $G(X)$ (possibly infinite) determined only implicitly by a kernel function to be chosen. Second, we formulate the constrained optimization problem of

$$minimize \ \frac{1}{2}\text{Tr}[\boldsymbol{EE}^{\text{T}}] + \frac{C}{2}\boldsymbol{U}^{\text{T}}\boldsymbol{U}$$
$$subject \ to \ \boldsymbol{T} - \boldsymbol{U}^{\text{T}}\boldsymbol{G}(\boldsymbol{X}) = \boldsymbol{E}.$$

Third, we make use of dual representations of the above constrained optimization problem to obtain $\boldsymbol{U} = \boldsymbol{G}^{\text{T}}\boldsymbol{a}$, where vector $\boldsymbol{a}$ takes the following form:

$$a = (C\boldsymbol{I} + \boldsymbol{K})^{-1}\boldsymbol{T}$$

and $\boldsymbol{K} = \boldsymbol{G}(\boldsymbol{X})\boldsymbol{G}^{\text{T}}(\boldsymbol{X})$ is a symmetric kernel matrix with elements $K_{nm} = g^{\text{T}}(x_n)g(x_m)$.

Finally, for each new input vector $\boldsymbol{x}$ in the test or dev set, we obtain the K-DSN (bottom) module's prediction as

$$y(x) = \boldsymbol{U}^{\text{T}}g(\boldsymbol{x}) = \mathbf{a}^{\text{T}}\boldsymbol{G}(\boldsymbol{X})g(\boldsymbol{x}) = \boldsymbol{k}^{\text{T}}(\boldsymbol{x})(C\boldsymbol{I} + \boldsymbol{K})^{-1}\boldsymbol{T},$$

where the kernel vector $\boldsymbol{k}(\boldsymbol{x})$ is so defined that its elements have values of $k_n(\boldsymbol{x}) = k(\boldsymbol{x}_n, \boldsymbol{x})$ in which $\boldsymbol{x}_n$ is a training sample and $\boldsymbol{x}$ is the current test sample.

For lth module in K-DCN where $l \geq 2$, the kernel matrix is modified to

$$\boldsymbol{K} = \boldsymbol{G}([\boldsymbol{X}|\boldsymbol{Y}^{(l-1)}|\,\boldsymbol{Y}^{(l-2)}|\dots\boldsymbol{Y}^{(1)}])\,\boldsymbol{G}^{\text{T}}([\boldsymbol{X}|\boldsymbol{Y}^{(l-1)}|\boldsymbol{Y}^{(l-2)}|\dots\boldsymbol{Y}^{(1)}]).$$

The key advantages of K-DSN can be analyzed as follows. First, unlike DSN which needs to compute hidden units' output, the K-DSN does not need to explicitly compute hidden units' output $\boldsymbol{G}(\boldsymbol{X})$ or $\boldsymbol{G}([\boldsymbol{X}|\boldsymbol{Y}^{(l-1)}|\boldsymbol{Y}^{(l-2)}|\dots\boldsymbol{Y}^{(1)}])$. When Gaussian kernels are used, kernel trick equivalently gives us an infinite number of hidden units without the need to compute them explicitly. Further, we no longer need to learn the lower-layer weight matrix $\boldsymbol{W}$ in DSN as described in [102] and the kernel parameter (e.g., the single variance parameter σ in the Gaussian kernel) makes K-DSN much less subject to overfitting than DSN. Figure 6.5 illustrates the basic architecture of a K-DSN using the Gaussian kernel and using three modules.

The entire K-DSN with Gaussian kernels is characterized by two sets of module-dependent hyper-parameters: $\sigma^{(l)}$ and $C^{(l)}$ the kernel

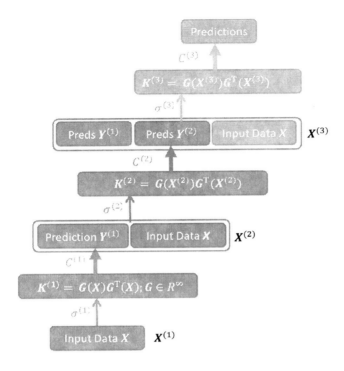

Figure 6.5: An example architecture of the K-DSN with three modules each of which uses a Gaussian kernel with different kernel parameters. [after [102], @IEEE].

smoothing parameter and regularization parameter, respectively. While both parameters are intuitive and their tuning (via line search or leave-one-out cross validation) is straightforward for a single bottom module, tuning the full network with all the modules is more difficult. For example, if the bottom module is tuned too well, then adding more modules would not benefit much. In contrast, when the lower modules are loosely tuned (i.e., relaxed from the results obtained from straightforward methods), the overall K-DSN often performs much better. The experimental results reported by Deng et al. [102] are obtained using a set of empirically determined tuning schedules to adaptively regularize the K-DSN from bottom to top modules.

The K-DSN described here has a set of highly desirable properties from the machine learning and pattern recognition perspectives. It combines the power of deep learning and kernel learning in a principled way and unlike the basic DSN there is no longer nonconvex optimization problem involved in training the K-DSN. The computation steps make the K-DSN easier to scale up for parallel computing in distributed servers than the DSN and tensor-DSN. There are many fewer parameters in the K-DSN to tune than in the DSN, T-DSN, and DNN, and there is no need for pre-training. It is found in the study of [102] that regularization plays a much more important role in the K-DSN than in the basic DSN and Tensor-DSN. Further, effective regularization schedules developed for learning the K-DSN weights can be motivated by intuitive insight from useful optimization tricks such as the heuristic in Rprop or resilient backpropagation algorithm [302].

However, as inherent in any kernel method, the scalability becomes an issue also for the K-DSN as the training and testing samples become very large. A solution is provided in the study by Huang et al. [171], based on the use of random Fourier features, which possess the strong theoretical property of approximating the Gaussian kernel while rendering efficient computation in both training and evaluation of the K-DSN with large training samples. It is empirically demonstrated that just like the conventional K-DSN exploiting rigorous Gaussian kernels, the use of random Fourier features also enables successful stacking of kernel modules to form a deep architecture.

7

Selected Applications in Speech and Audio Processing

7.1 Acoustic modeling for speech recognition

As discussed in Section 2, speech recognition is the very first success-ful application of deep learning methods at an industry scale. This success is a result of close academic-industrial collaboration, initiated at Microsoft Research, with the involved researchers identifying and acutely attending to the industrial need for large-scale deployment [68, 89, 109, 161, 323, 414]. It is also a result of carefully exploiting the strengths of the deep learning and the then-state-of-the-art speech recognition technology, including notably the highly efficient decoding techniques.

Speech recognition has long been dominated by the GMM–HMM method, with an underlying shallow or flat generative model of context-dependent GMMs and HMMs (e.g., [92, 93, 187, 293]). Neural networks once were a popular approach but had not been competitive with the GMM–HMM [42, 87, 261, 382]. Generative models with deep hidden dynamics likewise have also not been clearly competitive (e.g., [45, 73, 108, 282]).

Deep learning and the DNN started making their impact in speech recognition in 2010, after close collaborations between academic and industrial researchers; see reviews in [89, 161]. The collaborative work started in phone recognition tasks [89, 100, 135, 136, 257, 260, 258, 309, 311, 334], demonstrating the power of hybrid DNN architectures discussed in Section 5 and of subsequent new architectures with convolutional and recurrent structure. The work also showed the importance of raw speech features of spectrogram — back from the long-popular MFCC features toward but not yet reaching the raw speech-waveform level [183, 327]. The collaboration continued to large vocabulary tasks with more convincing, highly positive results [67, 68, 94, 89, 161, 199, 195, 223, 323, 353, 399, 414]. The success in large vocabulary speech recognition is in large part attributed to the use of a very large DNN output layer structured in the same way as the GMM–HMM speech units (senones), motivated partially by the speech researchers' desires to take advantage of the context-dependent phone modeling techniques that have been proven to work well in the GMM–HMM framework, and to keep the change of the already highly efficient decoder software's infrastructure developed for the GMM–HMM systems to a minimum. In the meantime, this body of work also demonstrated the possibility to reduce the need for the DBN-like pre-training in effective learning of DNNs when a large amount of labeled data is available. A combination of three factors helped to quickly spread the success of deep learning in speech recognition to the entire speech industry and academia: (1) significantly lowered errors compared with the then-state-of-the-art GMM-HMM systems; (2) minimal decoder changes required to deploy the new DNN-based speech recognizer due to the use of senones as the DNN output; and (3) reduced system complexity empowered by the DNN's strong modeling power. By the ICASSP-2013 timeframe, at least 15 major speech recognition groups worldwide confirmed experimentally the success of DNNs with very large tasks and with the use of raw speech spectral features other than MFCCs. The most notable groups include major industrial speech labs worldwide: Microsoft [49, 89, 94, 324, 399, 430], IBM [195, 309, 311, 307, 317], Google [69, 150, 184, 223], iFlyTek, and Baidu. Their results represent a new state-of-the-art in speech recognition widely deployed in these companies' voice products and services with extensive media coverage in recent years.

In the remainder of this chapter, we review a wide range of speech recognition work based on deep learning methods according to several major themes expressed in the section titles.

7.1.1 Back to primitive spectral features of speech

Deep learning, also referred as representation learning or (unsupervised) feature learning, sets an important goal of automatic discovery of powerful features from raw input data independent of application domains. For speech feature learning and for speech recognition, this goal is condensed to the use of primitive spectral or possibly waveform features. Over the past 30 years or so, largely "hand-crafted" transformations of speech spectrogram have led to significant accuracy improvements in the GMM-based HMM systems, despite the known loss of information from the raw speech data. The most successful transformation is the non-adaptive cosine transform, which gave rise to Mel-frequency cepstral coefficients (MFCC) features. The cosine transform approximately de-correlates feature components, which is important for the use of GMMs with diagonal covariance matrices. However, when GMMs are replaced by deep learning models such as DNNs, deep belief nets (DBNs), or deep autoencoders, such de-correlation becomes irrelevant due to the very strength of the deep learning methods in modeling data correlation. As discussed in detail in Section 4, early work of [100] demonstrated this strength and in particular the benefit of spectrograms over MFCCs in effective coding of bottleneck speech features using autoencoders in an unsupervised manner.

The pipeline from speech waveforms (raw speech features) to MFCCs and their temporal differences goes through intermediate stages of log-spectra and then (Mel-warped) filter-banks, with learned parameters based on the data. An important character of deep learning is to move away from separate design of feature representations and of classifiers. This idea of jointly learning classifier and feature transformation for speech recognition was already explored in early studies on the GMM–HMM based systems; e.g., [33, 50, 51, 299]. However, greater speech recognition performance gain is obtained only recently in the recognizers empowered by deep learning methods. For example, Mohamed et al. [259], Li et al. [221], and Deng et al. [94] showed significantly lowered speech recognition errors using large-scale DNNs when moving from the MFCC features back to more primitive (Mel-scaled) filter-bank features. These results indicate that DNNs can learn a better transformation than the original fixed cosine transform from the Mel-scaled filter-bank features.

Compared with MFCCs, "raw" spectral features not only retain more information, but also enable the use of convolution and pooling operations to represent and handle some typical speech variability — e.g., vocal tract length differences across speakers, distinct speaking styles causing formant undershoot or overshoot, etc. — expressed explicitly in the frequency domain. For example, the convolutional neural network (CNN) can only be meaningfully and effectively applied to speech recognition [1, 2, 3, 94] when spectral features, instead of MFCC features, are used.

More recently, Sainath et al. [307] went one step further toward raw features by learning the parameters that define the filter-banks on power spectra. That is, rather than using Mel-warped filter-bank features as the input features as in [1, 3, 50, 221], the weights corresponding to the Mel-scale filters are only used to initialize the parameters, which are subsequently learned together with the rest of the deep network as the classifier. The overall architecture of the jointly learned feature generator and classifier is shown in Figure 7.1. Substantial speech recognition error reduction is reported in [307].

It has been shown that not only learning the spectral aspect of the features are beneficial for speech recognition, learning the temporal aspect of the features is also helpful [332]. Further, Yu et al. [426] carefully analyzed the properties of different layers in the DNN as the layer-wise extracted features starting from the lower raw filter-bank features. They found that the improved speech recognition accuracy achieved by the DNNs partially attributes to DNN's ability to extract discriminative internal representations that are robust to the many sources of variability in speech signals. They also show that these representations become increasingly insensitive to small perturbations in

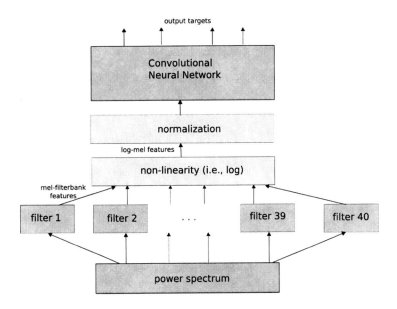

Figure 7.1: Illustration of the joint learning of filter parameters and the rest of the deep network. [after [307], @IEEE].

the input at higher layers, which helps to achieve better speech recognition accuracy.

To the extreme end, deep learning would promote to use the lowest level of raw features of speech, i.e., speech sound waveforms, for speech recognition, and learn the transformation automatically. As an initial attempt toward this goal the study carried out by Jaitly and Hinton [183] makes use of speech sound waves as the raw input feature to an RBM with a convolutional structure as the classifier. With the use of rectified linear units in the hidden layer [130], it is possible, to a limited extent, to automatically normalize the amplitude variation in the waveform signal. Although the final results are disappointing, the work shows that much work is needed along this direction. For example, just as demonstrated by Sainath et al. [307] that the use of raw spectra as features requires additional attention in normalization than MFCCs, the use of speech waveforms demands even more attention in normalization [327]. This is true for both GMM-based and deep learning based methods.

7.1.2   The DNN–HMM architecture versus use of DNN-derived features

Another major theme in the recent studies reported in the literature on applying deep learning methods to speech recognition is two disparate ways of using the DNN: (1) Direct applications of the DNN-HMM architecture as discussed in Section 5.3 to perform speech recognition; and (2) The use of DNNs to extract or derive features, which are then fed into a separate sequence classifier. In the speech recognition literature [42], a system, in which a neural network's output is directly used to estimate the emission probabilities of an HMM, is often called an ANN/HMM hybrid system. This should be distinguished from the use of "hybrid" in Section 5 and throughout this monograph, where a hybrid of unsupervised pre-training and of supervised fine tuning is exploited to learn the parameters of DNNs.

7.1.2.1   The DNN–HMM architecture as a recognizer

An early DNN–HMM architecture [257] was presented at the NIPS Workshop [109], developed, analyzed, and assisted by University of Toronto and MSR speech researchers. In this work, a five-layer DNN (called the DBN in the paper) was used to replace the Gaussian mixture models in the GMM–HMM system, and the monophone state was used as the modeling unit. Although monophones are generally accepted as a weaker phonetic representation than triphones, the DNN–HMM approach with monophones was shown to achieve higher phone recognition accuracy than the state-of-the-art triphone GMM–HMM systems. Further, the DNN results were found to be slightly superior to the then-best-performing single system based on the generative hidden trajectory model (HTM) in the literature [105, 108] evaluated on the same, commonly used TIMIT task by many speech researchers [107, 108, 274, 313]. At MSR, Redmond, the error patterns produced by these two separate systems (the DNN vs. the HTM) were carefully analyzed and found to be very different, reflecting distinct core capabilities of the two approaches and igniting intensive further studies on the DNN–HMM approach described below.

MSR and University of Toronto researchers [67, 68, 414] extended the DNN–HMM system from the monophone phonetic representation of the DNN outputs to the triphone or context-dependent counterpart and from phone recognition to large vocabulary speech recognition. Experiments conducted at MSR on the 24-hour and 48-hour Bing mobile voice search datasets collected under the real usage scenario demonstrate that the context-dependent DNN–HMM significantly outperforms the state-of-the-art GMM-HMM system. Three factors, in addition to the use of the DNN, contribute to the success: the use of tied triphones as the DNN modeling units, the use of the best available tri-phone GMM–HMM to generate the tri-phone state alignment, and the effective exploitation of a long window of input features. Experiments also indicate that the decoding time of a five-layer DNN–HMM is almost the same as that of the state-of-the-art triphone GMM–HMM.

The success was quickly extended to large vocabulary speech recognition tasks with hundreds and even thousands of hours of training set and with thousands of tri-phone states, including the Switchboard and Broadcast News databases, and Google's voice search and YouTube tasks [94, 161, 184, 309, 311, 324]. For example, on the Switchboard benchmark, the context-dependent DNN–HMM (CD-DNN–HMM) is shown to cut error by one third compared to the state-of-the-art GMM–HMM system [323]. As a summary, we show in Table 7.1 some quantitative recognition error rates in relatively early literature produced by the basic DNN–HMM architecture in comparison with those by the previous state-of-the-art systems based on the generative models. (More advanced architectures have produced better results than shown here). Note from sub-tables A to D, the training data are increased approximately one order of magnitude from one task to the next. Not only the computation scales up well (i.e., almost linearly) with the training size, but most importantly the relative error rate reduction increases substantially with increasing amounts of training data — from approximately 10% to 20%, and then to 30%. This set of results highlight the strongly desirable properties of the DNN-based methods, despite the conceptual simplicity of the overall DNN–HMM architecture and some known weaknesses.

Table 7.1: Comparisons of the DNN–HMM architecture with the generative model (e.g., the GMM–HMM) in terms of phone or word recognition error rates. From sub-tables A to D, the training data are increased approximately three orders of magnitudes.

Features	Setup	Error Rates
A: TIMIT Phone recognition (3 hours of training)		
GMM	w. Hidden dynamics	24.8%
DNN	5 layers × 2048	23.0%
B: Voice Search SER (24–48 hours of training)		
GMM	MPE (760 24-mix)	36.2%
DNN	5 layers × 2048	30.1%
C: Switch Board WER (309 hours of training)		
GMM	BMMI (9K 40-mix)	23.6%
DNN	7 layers × 2048	15.8%
D: Switch Board WER (2000 hours of training)		
GMM	BMMI (18K 72-mix)	21.7%
DNN	7 layers × 2048	14.6%

7.1.2.2 The use of DNN-derived features in a separate recognizer

One clear weakness of the above DNN–HMM architecture for speech recognition is that much of the highly effective techniques for the GMM–HMM systems, including discriminative training (in both feature space and model space), unsupervised speaker adaptation, noise robustness, and scalable batch training tools for big training data, developed over the past 20 some years may not be directly applicable to the new systems although similar techniques have been recently developed for DNN–HMMs. To remedy this problem, the "tandem" approach, developed originally by Hermansky et al. [154], has been adopted, where the output of the neural networks in the form of posterior probabilities for phone classes, are used, often in conjunction with the acoustic features to form new augmented input features, in a separate GMM–HMM system.

This tandem approach is used by Vinyals and Ravuri [379] where a DNN's outputs are extracted to serve as the features for mismatched noisy speech. It is reported that DNNs outperform the neural networks with a single hidden layer under the clean condition, but the gains slowly diminish as the noise level is increased. Furthermore, using MFCCs in conjunction with the posteriors computed from DNNs outperforms using the DNN features alone in low to moderate noise conditions with the tandem architecture. Comparisons of such tandem approach with the direct DNN–HMM approach are made by Tüske et al. [368] and Imseng et al. [182].

An alternative way of extracting the DNN features is to use the "bottleneck" layer, which is narrower than other layers in the DNN, to restrict the capacity of the network. Then, such bottleneck features are fed to a GMM–HMM system, often in conjunction with the original acoustic features and some dimensionality reduction techniques. The bottleneck features derived from the DNN are believed to capture information complementary to conventional acoustic features derived from the short-time spectra of the input. A speech recognizer based on the above bottleneck feature approach is built by Yu and Seltzer [425], with the overall architecture shown in Figure 7.2. Several variants of the DNN-based bottleneck-feature approach have been explored; see details in [16, 137, 201, 285, 308, 368].

Yet another method to derive the features from the DNN is to feed its top-most hidden layer as the new features for a separate speech

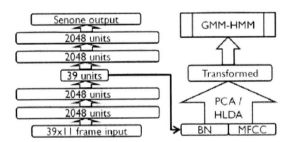

Figure 7.2: Illustration of the use of bottleneck (BN) features extracted from a DNN in a GMM–HMM speech recognizer. [after [425], @IEEE].

recognizer. In [399], a GMM–HMM is used as such a recognizer, and the high-dimensional, DNN-derived features are subject to dimensionality reduction before feeding them into the recognizer. More recently, a recurrent neural network (RNN) is used as the "backend" recognizer receiving the high-dimensional, DNN-derived features as the input without dimensionality reduction [48, 85]. These studies also show that the use of the top-most hidden layer of the DNN as features is better than other hidden layers and also better than the output layer in terms of recognition accuracy for the RNN sequence classifier.

7.1.3   Noise robustness by deep learning

The study of noise robustness in speech recognition has a long history, mostly before the recent rise of deep learning. One major contributing factor to the often observed brittleness of speech recognition technology is the inability of the standard GMM–HMM-based acoustic model to accurately model noise-distorted speech test data that differs in character from the training data, which may or may not be distorted by noise. A wide range of noise-robust techniques developed over past 30 years can be analyzed and categorized using five different criteria: (1) feature-domain versus model-domain processing, (2) the use of prior knowledge about the acoustic environment distortion, (3) the use of explicit environment-distortion models, (4) deterministic versus uncertainty processing, and (5) the use of acoustic models trained jointly with the same feature enhancement or model adaptation process used in the testing stage. See a comprehensive review in [220] and some additional review literature or original work in [4, 82, 119, 140, 230, 370, 404, 431, 444].

Many of the model-domain techniques developed for GMM–HMMs (e.g., model-domain noise robustness techniques surveyed by Li et al. [220] and Gales [119]) are not directly applicable to the new deep learning models for speech recognition. The feature-domain techniques, however, can be directly applied to the DNN system. A detailed investigation of the use of DNNs for noise robust speech recognition in the feature domain was reported by Seltzer et al. [325], who applied the C-MMSE [415] feature enhancement algorithm on the input feature used in the DNN. By processing both the training and testing data with the same algorithm, any consistent errors or artifacts introduced by the enhancement algorithm can be learned by the DNN–HMM recognizer. This study also successfully explored the use of the noise aware training paradigm for training the DNN, where each observation was augmented with an estimate of the noise. Strong results were obtained on the Aurora4 task. More recently, Kashiwagi et al. [191] applied the SPLICE feature enhancement technique [82] to a DNN speech recognizer. In that study the DNN's output layer was determined on clean data instead of on noisy data as in the study reported by Seltzer et al. [325].

Besides DNN, other deep architectures have also been proposed to perform feature enhancement and noise-robust speech recognition. For example, Mass et al. [235] applied a deep recurrent auto encoder neural network to remove noise in the input features for robust speech recognition. The model was trained on stereo (noisy and clean) speech features to predict clean features given noisy input, similar to the SPLICE setup but using a deep model instead of a GMM. Vinyals and Ravuri [379] investigated the tandem approaches to noise-robust speech recognition, where DNNs were trained directly with noisy speech to generate posterior features. Finally, Rennie et al. [300] explored the use of a version of the RBM, called the factorial hidden RBM, for noise-robust speech recognition.

7.1.4 Output representations in the DNN

Most deep learning methods for speech recognition and other information processing applications have focused on learning representations from input acoustic features without paying attention to output representations. The recent 2013 NIPS Workshop on Learning Output Representations (http://nips.cc/Conferences/2013/Program/event.php?ID=3714) was dedicated to bridging this gap. For example, the Deep Visual-Semantic Embedding Model described in [117], to be discussed more in Section 11) exploits continuous-valued output representations obtained from the text embeddings to assist in the branch of the deep network for classifying images. For speech recognition, the importance of designing effective linguistic representations for the output layers of deep networks is highlighted in [79].

Most current DNN systems use a high-dimensional output representation to match the context-dependent phonetic states in the HMMs. For this reason, the output layer evaluation can cost $1/3$ of the total computation time. To improve the decoding speed, techniques such as low-rank approximation is typically applied to the output layer. In [310] and [397], the DNN with high-dimensional output layer was trained first. The singular value decomposition (SVD)-based dimension reduction technique was then performed on the large output-layer matrix. The resulting matrices are further combined and as the result the original large weight matrix is approximated by a product of two much smaller matrices. This technique in essence converts the original large output layer to two layers — a bottleneck linear layer and a nonlinear output layer — both with smaller weight matrices. The converted DNN with reduced dimensionality is further refined. The experimental results show that no speech recognition accuracy reduction was observed even when the size is cut to half, while the run-time computation is significantly reduced.

The output representations for speech recognition can benefit from the structured design of the symbolic or phonological units of speech as presented in [79]. The rich phonological structure of symbolic nature in human speech has been well known for many years. Likewise, it has also been well understood for a long time that the use of phonetic or its finer state sequences, even with contextual dependency, in engineering speech recognition systems, is inadequate in representing such rich structure [86, 273, 355], and thus leaving a promising open direction to improve the speech recognition systems' performance. Basic theories about the internal structure of speech sounds and their relevance to speech recognition technology in terms of the specification, design, and learning of possible output representations of the underlying speech model for speech target sequences are surveyed in [76] and more recently in [79].

There has been a growing body of deep learning work in speech recognition with their focus placed on designing output representations related to linguistic structure. In [383, 384], a limitation of the output representation design, based on the context-dependent phone units as proposed in [67, 68], is recognized and a solution is offered. The root cause of this limitation is that all context-dependent phone states within a cluster created by the decision tree share the same set of parameters and this reduces its resolution power for fine-grained states during the decoding phase. The solution proposed formulates output representations of the context-dependent DNN as an instance of the canonical state modeling technique, making use of broad phonetic classes. First, triphones are clustered into multiple sets of shorter bi-phones using broad phone contexts. Then, the DNN is trained to discriminate the bi-phones within each set. Logistic regression is used to transform the canonical states into the detailed triphone state output probabilities. That is, the overall design of the output representation of the context-dependent DNN is hierarchical in nature, solving both the data sparseness and low-resolution problems at the same time.

Related work on designing the output linguistic representations for speech recognition can be found in [197] and in [241]. While the designs are in the context of GMM–HMM-based speech recognition systems, they both can be extended to deep learning models.

7.1.5 Adaptation of the DNN-based speech recognizers

The DNN–HMM is an advanced version of the artificial neural network and HMM "hybrid" system developed in 1990s, for which several adaptation techniques have been developed. Most of these techniques are based on linear transformation of the network weights of either input or output layers. A number of exploratory studies on DNN adaptation made use of the same or related linear transformation methods [223, 401, 402]. However, compared with the earlier narrower and shallower neural network systems, the DNN–HMM has significantly more parameters due to wider and deeper hidden layers used and the much larger output layer designed to model context dependent phones and states. This difference casts special challenges to adapting the DNN–HMM, especially when the adaptation data is small. Here we discuss representative recent studies on overcoming such challenges in adapting the large-sized DNN weights in distinct ways.

Yu et al. [430] proposed a regularized adaptation technique for DNNs. It adapts the DNN weights conservatively by forcing the distribution estimated from the adapted model to be close to that estimated from those before the adaptation. This constraint is realized by adding Kullback–Leibler divergence (KLD) regularization to the adaptation criterion. This type of regularization is shown to be equivalent to a modification of the target distribution in the conventional backpropagation algorithm and thus the training of the DNN remains largely unchanged. The new target distribution is derived to be a linear interpolation of the distribution estimated from the model before adaptation and the ground truth alignment of the adaptation data. This interpolation prevents overtraining by keeping the adapted model from straying too far from the speaker-independent model. This type of adaptation differs from L2 regularization, which constrains the model parameters themselves rather than the output probabilities.

In [330], adaptation of the DNN was applied not on the conventional network weights but on the hidden activation functions. In this way, the main limitation of current adaptation techniques based on adaptable linear transformation of the network weights in either the input or the output layer is effectively overcome, since the new method only needs to adapt a more limited number of hidden activation functions.

Several studies were carried out on unsupervised or semi-supervised adaptation of DNN acoustic models with different types of input features with success [223, 405].

Most recently, Saon et al. [317] explored a new and highly effective method in adapting DNNs for speech recognition. The method combined I-vector features with fMLLR (feature-domain max-likelihood linear regression) features as the input into a DNN. I-vectors or (speaker) identity vectors are commonly used for speaker verification and speaker recognition applications, as they encapsulate relevant information about a speaker's identity in a low-dimensional feature vector. The fMLLR is an effective adaptation technique developed for GMM–HMM systems. Since I-vectors do not obey locality in frequency, they must be combined carefully with the fMLLR features that obey locality. The architecture of the multi-scale CNN–DNN was shown to be effective for the combination of these two different types of features. During both training and decoding, the speaker-specific I-vector was appended to the frame-based fMLLR features.

7.1.6 Better architectures and nonlinear units

Over recent years, since the success of the (fully-connected) DNN–HMM hybrid system was demonstrated in [67, 68, 109, 161, 257, 258, 308, 309, 324, 429], many new architectures and nonlinear units have been proposed and evaluated for speech recognition. Here we provide an overview of this progress, extending the overview provided in [89].

The tensor version of the DNN is reported by Yu et al. [421, 422], which extends the conventional DNN by replacing one or more of its layers with a double-projection layer and a tensor layer. In the double-projection layer, each input vector is projected into two nonlinear sub-spaces. In the tensor layer, two subspace projections interact with each other and jointly predict the next layer in the overall deep architecture. An approach is developed to map the tensor layers to the conventional sigmoid layers so that the former can be treated and trained in a similar way to the latter. With this mapping the tensor version of the DNN can be treated as the DNN augmented with double-projection layers so that the backpropagation learning algorithm can be cleanly derived and relatively easily implemented.

A related architecture to the above is the tensor version of the DSN described in Section 6, also usefully applied to speech classification and recognition [180, 181]. The same approach applies to mapping the tensor layers (i.e., the upper layer in each of the many modules in the DSN context) to the conventional sigmoid layers. Again, this mapping simplifies the training algorithm so that it becomes not so far apart from that for the DSN.

As discussed in Section 3.2, the concept of convolution in time was originated in the TDNN (time-delay neural network) as a shallow neural network [202, 382] developed during early days of speech recognition. Only recently and when deep architectures (e.g. deep convolutional neural network or deep CNN) were used, it has been found that frequency-dimension weight sharing is more effective for high-performance phone recognition, when the HMM is used to handle the time variability, than time-domain weight sharing as in the previous TDNN in which the HMM was not used [1, 2, 3, 81]. These studies also show that designing the pooling scheme in the deep CNN to properly trade-off between invariance to vocal tract length and discrimination among speech sounds, together with a regularization technique of "dropout" [166], leads to even better phone recognition performance. This set of work further points to the direction of trading-off between trajectory discrimination and invariance expressed in the whole dynamic pattern of speech defined in mixed time and frequency domains using convolution and pooling. Moreover, the most recent studies reported in [306, 307, 312] show that CNNs also benefit large vocabulary continuous speech recognition. They further demonstrate that multiple convolutional layers provide even more improvement when the convolutional layers use a large number of convolution kernels or feature maps. In particular, Sainath et al. [306] extensively explored many variants of the deep CNN. In combination with several novel methods the deep CNN is shown to produce state of the art results in a few large vocabulary speech recognition tasks.

In addition to the DNN, CNN, and DSN, as well as their tensor versions, other deep models have also been developed and reported in the literature for speech recognition. For example, the deep-structured CRF, which stacks many layers of CRFs, have been usefully applied to the task of language identification [429], phone recognition [410], sequential labeling in natural language processing [428], and confidence calibration in speech recognition [423]. More recently, Demuynck and Triefenbach [70] developed the deep GMM architecture, where the aspects of DNNs that lead to strong performance are extracted and applied to build hierarchical GMMs. They show that by going "deep and wide" and feeding windowed probabilities of a lower layer of GMMs to a higher layer of GMMs, the performance of the deep-GMM system can be made comparable to a DNN. One advantage of staying in the GMM space is that the decades of work in GMM adaptation and discriminative learning remains applicable.

Perhaps the most notable deep architecture among all is the recurrent neural network (RNN) as well as its stacked or deep versions [135, 136, 153, 279, 377]. While the RNN saw its early success in phone recognition [304], it was not easy to duplicate due to the intricacy in training, let alone to scale up for larger speech recognition tasks. Learning algorithms for the RNN have been dramatically improved since then, and much better results have been obtained recently using the RNN [48, 134, 235], especially when the bi-directional LSTM (long short-term memory) is used [135, 136]. The basic information flow in the bi-directional RNN and a cell of LSTM is shown in Figures 7.3 and 7.4, respectively.

Learning the RNN parameters is known to be difficult due to vanishing or exploding gradients [280]. Chen and Deng [48] and Deng and

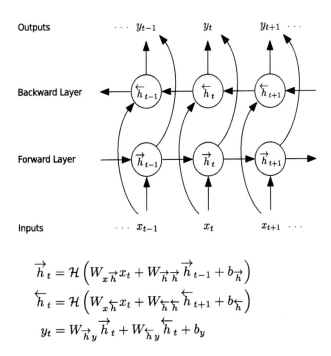

$$\overrightarrow{h}_t = \mathcal{H}\left(W_{x\overrightarrow{h}}x_t + W_{\overrightarrow{h}\overrightarrow{h}}\overrightarrow{h}_{t-1} + b_{\overrightarrow{h}}\right)$$

$$\overleftarrow{h}_t = \mathcal{H}\left(W_{x\overleftarrow{h}}x_t + W_{\overleftarrow{h}\overleftarrow{h}}\overleftarrow{h}_{t+1} + b_{\overleftarrow{h}}\right)$$

$$y_t = W_{\overrightarrow{h}y}\overrightarrow{h}_t + W_{\overleftarrow{h}y}\overleftarrow{h}_t + b_y$$

Figure 7.3: Information flow in the bi-directional RNN, with both diagrammatic and mathematical descriptions. W's are weight matrices, not shown but can be easily inferred in the diagram. [after [136], @IEEE].

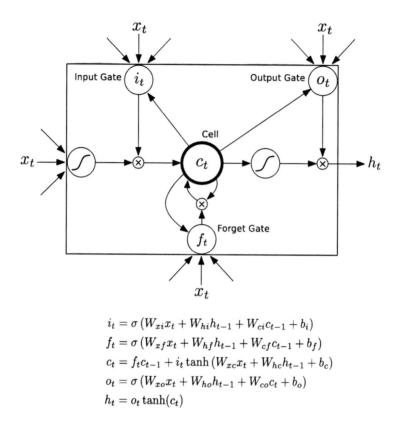

$$i_t = \sigma \left(W_{xi} x_t + W_{hi} h_{t-1} + W_{ci} c_{t-1} + b_i \right)$$
$$f_t = \sigma \left(W_{xf} x_t + W_{hf} h_{t-1} + W_{cf} c_{t-1} + b_f \right)$$
$$c_t = f_t c_{t-1} + i_t \tanh \left(W_{xc} x_t + W_{hc} h_{t-1} + b_c \right)$$
$$o_t = \sigma \left(W_{xo} x_t + W_{ho} h_{t-1} + W_{co} c_t + b_o \right)$$
$$h_t = o_t \tanh(c_t)$$

Figure 7.4: Information flow in an LSTM unit of the RNN, with both diagrammatic and mathematical descriptions. W's are weight matrices, not shown but can easily be inferred in the diagram. [after [136], @IEEE].

Chen [85] developed a primal-dual training method that formulates the learning of the RNN as a formal optimization problem, where cross entropy is maximized subject to the condition that the infinity norm of the recurrent matrix of the RNN is less than a fixed value to guarantee the stability of RNN dynamics. Experimental results on phone recognition demonstrate: (1) the primal-dual technique is highly effective in learning RNNs, with superior performance to the earlier heuristic method of truncating the size of the gradient; (2) The use of a DNN to compute high-level features of speech data to feed into the RNN gives much higher accuracy than without using the DNN; and (3) The accuracy drops progressively as the DNN features are extracted from higher to lower hidden layers of the DNN.

A special case of the RNN is reservoir models or echo state networks, where the output layers are fixed to be linear instead of nonlinear as in the regular RNN, and where the recurrent matrices are carefully designed but not learned. The input matrices are also fixed and not learned, due partly to the difficulty of learning. Only the weight matrices between the hidden and output layers are learned. Since the output layer is linear, the learning is very efficient and with global optimum achievable by a closed-form solution. But due to the fact that many parameters are not learned, the hidden layer needs to be very large in order to obtain good results. Triefenbach et al. [365] applied such models to phone recognition, with reasonably good accuracy obtained.

Palangi et al. [276] presented an improved version of the reservoir model by learning both the input and recurrent matrices which were fixed in the previous model that makes use of the linear output (or readout) units to simplify the learning of only the output matrix in the RNN. Rather, a special technique is devised that takes advantage of the linearity in the output units in the reservoir model to learn the input and recurrent matrices. Compared with the backpropagation through time (BPTT) algorithm commonly used in learning the general RNNs, the proposed technique makes use of the linearity in the output units to provide constraints among various matrices in the RNN, enabling the computation of the gradients as the learning signal in an analytical form instead of by recursion as in the BPTT.

In addition to the recent innovations in better architectures of deep learning models for speech recognition reviewed above, there is also a growing body of work on developing and implementing better nonlinear units. Although sigmoidal and tanh functions are the most commonly used nonlinear types in DNNs their limitations are well known. For example, it is slow to learn the whole network due to weak gradients when the units are close to saturation in both directions. Jaitly and Hinton [183] appear to be the first to apply the rectified linear units (ReLU) in the DNNs to speech recognition to overcome the weakness of the sigmoidal units. ReLU refers to the units in a neural network that use the activation function of $f(x) = \max(0, x)$. Dahl et al. [65]

and Mass et al. [234] successfully applied ReLU to large vocabulary speech recognition, with the best accuracy obtained when combining ReLU with the "Dropout" regularization technique.

Another new type of DNN units demonstrated more recently to be useful for speech recognition is the "maxout" units, which were used for forming the deep maxout network as described in [244]. A deep maxout network consists of multiple layers which generate hidden activations via the maximum or "maxout" operation over a fixed number of weighted inputs called a "group." This is the same operation as the max pooling used in the CNN as discussed earlier for both speech recognition and computer vision. The maximal value within each group is taken as the output from the previous layer. Most recently, Zhang et al. [441] generalize the above "maxout" units to two new types. The "soft-maxout" type of units replace the original max operation with the soft-max function. The second, p-norm type of units used the nonlinearity of $y = \|x\|_p$. It is shown experimentally that the p-norm units with $p = 2$ perform consistently better than the maxout, tanh, and ReLU units. In Gulcehre et al. [138], techniques that automatically learn the p-norm was proposed and investigated.

Finally, Srivastava et al. [350] propose yet another new type of nonlinear units, called winner-take-all units. Here, local competition among neighboring neurons are incorporated into the otherwise regular feedforward architecture, which is then trained via backpropagation with different gradients than the normal one. Winner-take-all is an interesting new form of nonlinearity, and it forms groups of (typically two) neurons where all the neurons in a group are made zero-valued except the one with the largest value. Experiments show that the network does not forget as much as networks with standard sigmoidal nonlinearity. This new type of nonlinear units are yet to be evaluated in speech recognition tasks.

7.1.7 Better optimization and regularization

Another area where significant advances are made recently in applying deep learning to acoustic model for speech recognition is on optimization criteria and methods, as well as on the related regularization techniques to help prevent overfitting during the deep network training.

One of the early studies on DNNs for speech recognition, conducted at Microsoft Research and reported in [260], first recognizes the mismatch between the desired error rate and the cross-entropy training criterion in the conventional DNN training. The solution is provided by replacing the frame-based, cross-entropy training criterion with the full-sequence-based maximum mutual information optimization objective, in a similar way to defining the training objective for the shallow neural network interfaced with an HMM [194]. Equivalently, this amounts to putting the model of conditional random field (CRF) at the top of the DNN, replacing the original softmax layer which naturally leads to cross entropy. (Note the DNN was called the DBN in the paper). This new sequential discriminative learning technique is developed to jointly optimize the DNN weights, CRF transition weights, and bi-phone language model. Importantly, the speech task is defined in TIMIT, with the use of a simple bi-phone-gram "language" model. The simplicity of the bi-gram language model enables the full-sequence training to carry out without the need to use lattices, drastically reducing the training complexity.

As another way to motivate the full-sequence training method of [260], we note that the earlier DNN phone recognition experiments made use of the standard frame-based objective function in static pattern classification, cross-entropy, to optimize the DNN weights. The transition parameters and language model scores were obtained from an HMM and were trained independently of the DNN weights. However, it has been known during the long history of the HMM research that sequence classification criteria can be very helpful in improving speech and phone recognition accuracy. This is because the sequence classification criteria are more directly correlated with the performance measure (e.g., the overall word or phone error rate) than frame-level criteria. More specifically, the use of frame-level cross entropy to train the DNN for phone sequence recognition does not explicitly take into account the fact that the neighboring frames have smaller distances between the assigned probability distributions over phone class labels. To overcome this deficiency, one can optimize the conditional probability of the whole sequence of labels, given the whole visible feature utterance or equivalently the hidden feature sequence extracted by DNN. To optimize the log conditional probability on the training data, the gradient can be taken over the activation parameters, transition parameters and lower-layer weights, and then pursue back-propagation of the error defined at the sentence level. We remark that in a much earlier study [212], combining a neural network with a CRF-like structure was done, where the mathematical formulation appears to include CRFs as a special case. Also, the benefit of using the full-sequence classification criteria was shown earlier on shallow neural networks in [194, 291].

In implementing the above full-sequence learning algorithm for the DNN system as described in [260], the DNN weights are initialized using the frame-level cross entropy as the objective. The transition parameters are initialized from the combination of the HMM transition matrices and the "bi-phone language" model scores, and are then further optimized by tuning the transition features while fixing the DNN weights before the joint optimization. Using joint optimization with careful scheduling to reduce overfitting, it is shown that the full-sequence training outperforms the DNN trained with frame-level cross entropy by approximately 5% relative [260]. Without the effort to reduce overfitting, it is found that the DNN trained with MMI is much more prone to overfitting than that trained with frame-level cross entropy. This is because the correlations across frames in speech tend to be different among the training, development, and test data. Importantly, such differences do not show when frame-based objective functions are used for training.

For large vocabulary speech recognition where more complex language models are in use, the optimization methods for full-sequence training of the DNN–HMM are much more sophisticated. Kingsbury et al. [195] reported the first success of such training using parallel, second-order, Hessian-free optimization techniques, which are carefully implemented for large vocabulary speech recognition. Sainath et al. [305] improved and speeded up the Hessian-free techniques by reducing the number of Krylov subspace solver iterations [378], which are used for implicit estimation of the Hessian. They also use sampling methods to decrease the amount of training data to speed up the training. While the batch-mode, second-order Hessian-free techniques prove successful for full-sequence training of large-scale DNN–HMM systems, the success of the first-order stochastic gradient descent methods is also reported recently [353]. It is found that heuristics are needed to handle the problem of lattice sparseness. That is, the DNN must be adjusted to the updated numerator lattices by additional iterations of frame-based cross-entropy training. Further, artificial silence arcs need to be added to the denominator lattices, or the maximum mutual information objective function needs to be smoothed with the frame-based cross entropy objective. The conclusion is that for large vocabulary speech recognition tasks with sparse lattices, the implementation of the sequence training requires much greater engineering skills than the small tasks such as reported in [260], although the objective function as well as the gradient derivation are essentially the same. Similar conclusions are reached by Vesely et al. [374] when carrying out full-sequence training of DNN–HMMs for large-vocabulary speech recognition. However, different heuristics from [353] are shown to be effective in the training. Separately, Wiesler et al. [390] investigated the Hessian-free optimization method for training the DNN with the cross-entropy objective and empirically analyzed the properties of the method. And finally, Dognin and Goel [113] combined stochastic average gradient and Hessian-free optimization for sequence training of deep neural networks with success in that the training procedure converges in about half the time compared with the full Hessian-free sequence training.

For large DNN–HMM systems with either frame-level or sequence-level optimization objectives, speeding up the training is essential to take advantage of large amounts of training data and of large model sizes. In addition to the methods described above, Dean et al. [69] reported the use of the asynchronous stochastic gradient descent (ASGD) method, the adaptive gradient descent (Adagrad) method, and the large-scale limited-memory BFGS (L-BFGS) method for very large vocabulary speech recognition. Sainath et al. [312] provided a review of a wide range of optimization methods for speeding up the training of DNN-based systems for large speech recognition tasks.

In addition to the advances described above focusing on optimization with the fully supervised learning paradigm, where all training data contain the label information, the semi-supervised training paradigm is also exploited for learning DNN–HMM systems for speech recognition. Liao et al. [223] reported the exploration of using semi-supervised training on the DNN–HMM system for the very challenging task of recognizing YouTube speech. The main technique is based on the use of "island of confidence" filtering heuristics to select useful training segments. Separately, semi-supervised training of DNNs is explored by Vesely et al. [374], where self-training strategies are used as the basis for data selection using both the utterance-level and frame-level confidences. Frame-selection based on per-frame confidences derived from confusion in a lattice is found beneficial. Huang et al. [176] reported another variant of semi-supervised training technique in which multi-system combination and confidence recalibration is applied to select the training data. Further, Thomas et al. [362] overcome the problem of lacking sufficient training data for acoustic modeling in a number of low-resource scenarios. They make use of transcribed multilingual data and semi-supervised training to build the proposed feature front-ends for subsequent speech recognition.

Finally, we see important progress in deep learning based speech recognition in recent years with the introduction of new regularization methods based on "dropout" originally proposed by Hinton et al. [166]. Overfitting is very common in DNN training and co-adaptation is prevalent within the DNN with multiple activations adapting together to explain input acoustic data. Dropout is a technique to limit co-adaptation. It operates as follows. On each training instance, each hidden unit is randomly omitted with a fixed probability (e.g., $p = 0.5$). Then, decoding is done normally except with straightforward scaling of the DNN weights (by a factor of $1 - p$). Alternatively, the scaling of the DNN weights can be done during training [by a factor of $1/(1-p)$] rather than in decoding. The benefits of dropout regularization for training DNNs are to make a hidden unit in the DNN act strongly by itself without relying on others, and to serve a way to do model averaging of different networks. These benefits are most pronounced when the training data is limited, or when the DNN size is disproportionally large with respect to the size of the training data. Dahl et al. [65] applied dropout in conjunction with the ReLU units and to only the top few layers of a fully-connected DNN. Seltzer and Yu [325] applied it to noise robust speech recognition. Deng et al. [81], on the other hand, applied dropout to all layers of a deep convolutional neural network, including both the top fully connected DNN layers and the bottom locally connected CNN layer and the pooling layer. It is found that the dropout rate need to be substantially smaller for the convolutional layer.

Subsequent work on applying dropout includes the study by Miao and Metze [243], where DNN-based speech recognition is constrained by low resources with sparse training data. Most recently, Sainath et al. [306] combined dropout with a number of novel techniques described in this section (including the use of deep CNNs, Hessian-free sequence learning, the use of ReLU units, and the use of joint fMLLR and filter-bank features, etc.) to obtain state of the art results on several large vocabulary speech recognition tasks.

As a summary, the initial success of deep learning methods for speech analysis and recognition reported around 2010 has come a long way over the past three years. An explosive growth in the work and publications on this topic has been observed, and huge excitement has been ignited within the speech recognition community. We expect that the growth in the research on deep learning based speech recognition will continue, at least in the near future. It is also fair to say that the continuing large-scale success of deep learning in speech recognition as surveyed in this chapter (up to the ASRU-2013 time frame) is a key stimulant to the large-scale exploration and applications of the deep learning methods to other areas, which we will survey in Sections 8–11.

7.2 Speech synthesis

In addition to speech recognition, the impact of deep learning has recently spread to speech synthesis, aimed to overcome the limitations of the conventional approach in statistical parametric synthesis based on Gaussian-HMM and decision-tree-based model clustering. The goal of speech synthesis is to generate speech sounds directly from text and possibly with additional information. The first set of papers appeared at ICASSP, May 2013, where four different deep learning approaches are reported to improve the traditional HMM-based statistical parametric speech synthesis systems built based on "shallow" speech models, which we briefly review here after providing appropriate background information.

Statistical parametric speech synthesis emerged in the mid-1990s, and is currently the dominant technology in speech synthesis. See a recent overview in [364]. In this approach, the relationship between texts and their acoustic realizations are modeled using a set of stochastic generative acoustic models. Decision tree-clustered context-dependent HMMs with a Gaussian distribution as the output of an HMM state are the most popular generative acoustic model used. In such HMM-based speech synthesis systems, acoustic features including the spectra, excitation and segment durations of speech are modeled simultaneously within a unified context-dependent HMM framework. At the synthesis time, a text analysis module extracts a sequence of contextual factors including phonetic, prosodic, linguistic, and grammatical descriptions from an input text to be synthesized. Given the sequence of contextual factors, a sentence-level context-dependent HMM corresponding to the input text is composed, where its model parameters are determined by traversing the decision trees. The acoustic features are predicted so as to maximize their output probabilities from the sentence HMM under the constraints between static and dynamic features. Finally, the predicted acoustic features are sent to a waveform synthesis module to reconstruct the speech waveforms. It has been known for many years that the speech sounds generated by this standard approach are often muffled compared with natural speech. The inadequacy of acoustic modeling based on the shallow-structured HMM is conjectured to be one of the reasons. Several very recent studies have adopted deep learning approaches to overcome such deficiency. One significant advantage of deep learning techniques is their strong ability to represent the intrinsic correlation or mapping relationship among the units of a high-dimensional stochastic vector using a generative (e.g., the RBM and DBN discussed in Section 3.2) or discriminative (e.g., the DNN discussed in Section 3.3) modeling framework. The deep learning techniques are thus expected to help the acoustic modeling aspect of speech synthesis in overcoming the limitations of the conventional shallow modeling approach.

A series of studies are carried out recently on ways of overcoming the above limitations using deep learning methods, inspired partly by the intrinsically hierarchical processes in human speech production and the successful applications of a number of deep learning methods in speech recognition as reviewed earlier in this chapter. In Ling et al. [227, 229], the RBM and DBN as generative models are used to replace the traditional Gaussian models, achieving significant quality improvement, in both subjective and objective measures, of the synthesized voice. In the approach developed in [190], the DBN as a generative model is used to represent joint distribution of linguistic and acoustic features. Both the decision trees and Gaussian models are replaced by the DBN. The method is very similar to that used for generating digit images by the DBN, where the issue of temporal sequence modeling specific to speech (non-issue for image) is by-passed via the use of the relatively large, syllable-sized units in speech synthesis. On the other hand, in contrast to the generative deep models (RBMs and DBNs) exploited above, the study reported in [435] makes use of the discriminative model of the DNN to represent the conditional distribution of the acoustic features given the linguistic features. Finally, in [115], the discriminative model of the DNN is used as a feature extractor that summarizes high-level structure from the raw acoustic features. Such DNN features are then used as the input for the second stage for the prediction of prosodic contour targets from contextual features in the full speech synthesis system.

The application of deep learning to speech synthesis is in its infancy, and much more work is expected from that community in the near future.

7.3 Audio and music processing

Similar to speech recognition but to a less extent, in the area of audio and music processing, deep learning has also become of intense interest but only quite recently. As an example, the first major event of deep learning for speech recognition took place in 2009, followed by a series of events including a comprehensive tutorial on the topic at ICASSP-2012 and with the special issue at IEEE Transactions on Audio, Speech, and Language Processing, the premier publication for speech recognition, in the same year. The first major event of deep learning for audio and music processing appears to be the special session at ICASSP-2014, titled Deep Learning for Music [14].

In the general field of audio and music processing, the impacted areas by deep learning include mainly music signal processing and music information retrieval [15, 22, 141, 177, 178, 179, 319]. Deep learning presents a unique set of challenges in these areas. Music audio signals are time series where events are organized in musical time, rather than in real time, which changes as a function of rhythm and expression. The measured signals typically combine multiple voices that are synchronized in time and overlapping in frequency, mixing both short-term and long-term temporal dependencies. The influencing factors include musical tradition, style, composer and interpretation. The high complexity and variety give rise to the signal representation problems well-suited to the high levels of abstraction afforded by the perceptually and biologically motivated processing techniques of deep learning.

In the early work on audio signals as reported by Lee et al. [215] and their follow-up work, the convolutional structure is imposed on the RBM while building up a DBN. Convolution is made in time by sharing weights between hidden units in an attempt to detect the same "invariant" feature over different times. Then a max-pooling operation is performed where the maximal activations over small temporal neighborhoods of hidden units are obtained, inducing some local temporal invariance. The resulting convolutional DBN is applied to audio as well as speech data for a number of tasks including music artist and genre classification, speaker identification, speaker gender classification, and phone classification, with promising results presented.

The RNN has also been recently applied to music processing applications [22, 40, 41], where the use of ReLU hidden units instead of logistic or tanh nonlinearities are explored in the RNN. As reviewed in

Section 7.2, ReLU units compute $y = \max(x, 0)$, and lead to sparser gradients, less diffusion of credit and blame in the RNN, and faster training. The RNN is applied to the task of automatic recognition of chords from audio music, an active area of research in music information retrieval. The motivation of using the RNN architecture is its power in modeling dynamical systems. The RNN incorporates an internal memory, or hidden state, represented by a self-connected hidden layer of neurons. This property makes them well suited to model temporal sequences, such as frames in a magnitude spectrogram or chord labels in a harmonic progression. When well trained, the RNN is endowed with the power to predict the output at the next time step given the previous ones. Experimental results show that the RNN-based automatic chord recognition system is competitive with existing state-of-the-art approaches [275]. The RNN is capable of learning basic musical properties such as temporal continuity, harmony and temporal dynamics. It can also efficiently search for the most musically plausible chord sequences when the audio signal is ambiguous, noisy or weakly discriminative.

A recent review article by Humphrey et al. [179] provides a detailed analysis on content-based music informatics, and in particular on why the progress is decelerating throughout the field. The analysis concludes that hand-crafted feature design is sub-optimal and unsustainable, that the power of shallow architectures is fundamentally limited, and that short-time analysis cannot encode musically meaningful structure. These conclusions motivate the use of deep learning methods aimed at automatic feature learning. By embracing feature learning, it becomes possible to optimize a music retrieval system's internal feature representation or discovering it directly, since deep architectures are especially well-suited to characterize the hierarchical nature of music. Finally, we review the very recent work by van den Oord, et al. [371] on content-based music recommendation using deep learning methods. Automatic music recommendation has become an increasingly significant and useful technique in practice. Most recommender systems rely on collaborative filtering, suffering from the cold start problem where it fails when no usage data is available. Thus, collaborative filtering is not effective for recommending new and unpopular songs. Deep learning methods power the latent factor model for recommendation, which predicts the latent factors from music audio when they cannot be obtained from usage data. A traditional approach using a bag-of-words representation of the audio signals is compared with deep CNNs with rigorous evaluation made. The results show highly sensible recommendations produced by the predicted latent factors using deep CNNs. The study demonstrates that a combination of convolutional neural networks and richer audio features lead to such promising results for content-based music recommendation.

Like speech recognition and speech synthesis, much more work is expected from the music and audio signal processing community in the near future.

8

Selected Applications in Language Modeling and Natural Language Processing

Research in language, document, and text processing has seen increasing popularity recently in the signal processing community, and has been designated as one of the main focus areas by the IEEE Signal Processing Society's Speech and Language Processing Technical Committee. Applications of deep learning to this area started with language modeling (LM), where the goal is to provide a probability to any arbitrary sequence of words or other linguistic symbols (e.g., letters, characters, phones, etc.). Natural language processing (NLP) or computational linguistics also deals with sequences of words or other linguistic symbols, but the tasks are much more diverse (e.g., translation, parsing, text classification, etc.), not focusing on providing probabilities for linguistic symbols. The connection is that LM is often an important and very useful component of NLP systems. Applications to NLP is currently one of the most active areas in deep learning research, and deep learning is also considered as one promising direction by the NLP research community. However, the intersection between the deep learning and NLP researchers is so far not nearly as large as that for the application areas of speech or vision. This is partly because the hard evidence for the superiority of deep learning over the current state of the art NLP methods has not been as strong as speech or visual object recognition.

8.1 Language modeling

Language models (LMs) are crucial part of many successful applications, such as speech recognition, text information retrieval, statistical machine translation and other tasks of NLP. Traditional techniques for estimating the parameters in LMs are based on N-gram counts. Despite known weaknesses of N-grams and huge efforts of research communities across many fields, N-grams remained the state-of-the-art until neural network and deep learning based methods were shown to significantly lower the perplexity of LMs, one common (but not ultimate) measure of the LM quality, over several standard benchmark tasks [245, 247, 248].

Before we discuss neural network based LMs, we note the use of hierarchical Bayesian priors in building up deep and recursive structure for LMs [174]. Specifically, Pitman-Yor process is exploited as the Bayesian prior, from which a deep (four layers) probabilistic generative model is built. It offers a principled approach to LM smoothing by incorporating the power-law distribution for natural language. As discussed in Section 3, this type of prior knowledge embedding is more readily achievable in the generative probabilistic modeling setup than in the discriminative neural network based setup. The reported results on LM perplexity reduction are not nearly as strong as that achieved by the neural network based LMs, which we discuss next.

There has been a long history [19, 26, 27, 433] of using (shallow) feed-forward neural networks in LMs, called the NNLM. The use of DNNs in the same way for LMs appeared more recently in [8]. An LM is a function that captures the salient statistical characteristics of the distribution of sequences of words in natural language. It allows one to make probabilistic predictions of the next word given preceding ones. An NNLM is one that exploits the neural network's ability to learn distributed representations in order to reduce the impact of the curse of dimensionality. The original NNLM, with a feed-forward neural network structure works as follows: the input of the N-gram NNLM is formed by using a fixed length history of $N - 1$ words. Each of the previous $N - 1$ words is encoded using the very sparse 1-of-V coding, where V is the size of the vocabulary. Then, this 1-of-V orthogonal representation of words is projected linearly to a lower dimensional space, using the projection matrix shared among words at different positions in the history. This type of continuous-space, distributed representation of words is called "word embedding," very different from the common symbolic or localist presentation [26, 27]. After the projection layer, a hidden layer with nonlinear activation function, which is either a hyperbolic tangent or a logistic sigmoid, is used. An output layer of the neural network then follows the hidden layer, with the number of output units equal to the size of the full vocabulary. After the network is trained, the output layer activations represent the "N-gram" LM's probability distribution.

The main advantage of NNLMs over the traditional counting-based N-gram LMs is that history is no longer seen as exact sequence of $N - 1$ words, but rather as a projection of the entire history into some lower dimensional space. This leads to a reduction of the total number of parameters in the model that have to be trained, resulting in automatic clustering of similar histories. Compared with the class-based N-gram LMs, the NNLMs are different in that they project all words into the same low dimensional space, in which there can be many degrees of similarity between words. On the other hand, NNLMs have much larger computational complexity than N-gram LMs.

Let's look at the strengths of the NNLMs again from the viewpoint of distributed representations. A distributed representation of a symbol is a vector of features which characterize the meaning of the symbol. Each element in the vector participates in representing the meaning. With an NNLM, one relies on the learning algorithm to discover meaningful, continuous-valued features. The basic idea is to learn to associate each word in the dictionary with a continuous-valued vector representation, which in the literature is called a word embedding, where each word corresponds to a point in a feature space. One can imagine that each dimension of that space corresponds to a semantic or grammatical characteristic of words. The hope is that functionally similar words get to be closer to each other in that space, at least along some directions. A sequence of words can thus be transformed into a sequence of these learned feature vectors. The neural network learns to map that sequence of feature vectors to the probability distribution over the next word in the sequence. The distributed representation approach to LMs has the advantage that it allows the model to generalize well to sequences that are not in the set of training word sequences, but that are similar in terms of their features, i.e., their distributed representation. Because neural networks tend to map nearby inputs to nearby outputs, the predictions corresponding to word sequences with similar features are mapped to similar predictions.

The above ideas of NNLMs have been implemented in various studies, some involving deep architectures. The idea of structuring hierarchically the output of an NNLM in order to handle large vocabularies was introduced in [18, 262]. In [252], the temporally factored RBM was used for language modeling. Unlike the traditional N-gram model, the factored RBM uses distributed representations not only for context words but also for the words being predicted. This approach is generalized to deeper structures as reported in [253]. Subsequent work on NNLM with "deep" architectures can be found in [205, 207, 208, 245, 247, 248]. As an example, Le et al. [207] describes an NNLM with structured output layer (SOUL–NNLM) where the processing depth in the LM is focused in the neural network's output representation. Figure 8.1 illustrates the SOUL-NNLM architecture with hierarchical structure in the output layers of the neural network, which shares the same architecture with the conventional NNLM up to the hidden layer. The hierarchical structure for the network's output vocabulary is in the form of a clustering tree, shown to the right of Figure 8.1, where each word belongs to only one class and ends in a single leaf node of the tree. As a result of the hierarchical structure, the SOUL–NNLM enables the training of the NNLM with a full, very large vocabulary. This gives advantages over the traditional NNLM which requires shortlists of words in order to carry out the efficient computation in training.

As another example neural-network-based LMs, the work described in [247, 248] and [245] makes use of RNNs to build large scale language

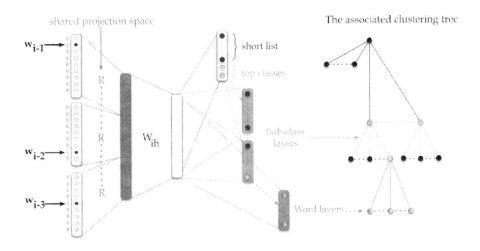

Figure 8.1: The SOUL–NNLM architecture with hierarchical structure in the output layers of the neural network [after [207], @IEEE].

models, called RNNLMs. The main difference between the feed-forward and the recurrent architecture for LMs is different ways of representing the word history. For feed-forward NNLM, the history is still just previous several words. But for the RNNLM, an effective representation of history is learned from the data during training. The hidden layer of RNN represents all previous history and not just $N - 1$ previous words, thus the model can theoretically represent long context patterns. A further important advantage of the RNNLM over the feed-forward counterpart is the possibility to represent more advanced patterns in the word sequence. For example, patterns that rely on words that could have occurred at variable positions in the history can be encoded much more efficiently with the recurrent architecture. That is, the RNNLM can simply remember some specific word in the state of the hidden layer, while the feed-forward NNLM would need to use parameters for each specific position of the word in the history.

The RNNLM is trained using the algorithm of back-propagation through time; see details in [245], which provided Figure 8.2 to show during training how the RNN unfolds as a deep feed-forward network (with three time steps back in time).

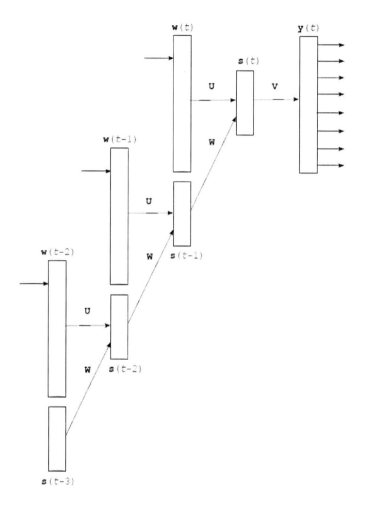

Figure 8.2: During the training of RNNLMs, the RNN unfolds into a deep feed-forward network; based on Figure 3.2 of [245].

The training of the RNNLM achieves stability and fast convergence, helped by capping the growing gradient in training RNNs. Adaptation schemes for the RNNLM are also developed by sorting the training data with respect to their relevance and by training the model during processing of the test data. Empirical comparisons with other state-of-the-art counting-based N-gram LMs show much better performance of RNNLM in the perplexity measure, as reported in [247, 248] and [245].

A separate work on applying RNN to an LM with the unit of characters instead of words can be found in [153, 357]. Many interesting properties such as predicting long-term dependencies (e.g., making open and closing quotes in a paragraph) are demonstrated. However, the usefulness of characters instead of words as units in practical applications is not clear because the word is such a powerful representation for natural language. Changing words to characters in LMs may limit most practical application scenarios and the training become more difficult. Word-level models currently remain superior.

In the most recent work, Mnih and Teh [255] and Mnih and Kavukcuoglu [254] have developed a fast and simple training algorithm for NNLMs. Despite their superior performance, NNLMs have been used less widely than standard N-gram LMs due to the much longer training time. The reported algorithm makes use of a method called noise-contrastive estimation or NCE [139] to achieve much faster training for NNLMs, with time complexity independent of the vocabulary size; hence a flat instead of tree-structured output layer in the NNLM is used. The idea behind NCE is to perform nonlinear logistic regression to discriminate between the observed data and some artificially generated noise. That is, to estimate parameters in a density model of observed data, we can learn to discriminate between samples from the data distribution and samples from a known noise distribution. As an important special case, NCE is particularly attractive for unnormalized distributions (i.e., free from partition functions in the denominator). In order to apply NCE to train NNLMs efficiently, Mnih and Teh [255] and Mnih and Kavukcuoglu [254] first formulate the learning problem as one which takes the objective function as the distribution of the word in terms of a scoring function. The NNLM then can be viewed as a way to quantify the compatibility between the word history and a candidate next word using the scoring function. The objective function for training the NNLM thus becomes exponentiation of the scoring function, normalized by the same constant over all possible words. Removing the costly normalization factor, NCE is shown to speed up the NNLM training over an order of magnitude.

A similar concept to NCE is used in the recent work of [250], which is called negative sampling. This is applied to a simplified version of an NNLM, for the purpose of constructing word embedding instead of computing probabilities of word sequences. Word embedding is an important concept for NLP applications, which we discuss next.

8.2 Natural language processing

Machine learning has been a dominant tool in NLP for many years. However, the use of machine learning in NLP has been mostly limited to numerical optimization of weights for human designed representations and features from the text data. The goal of deep or representation learning is to automatically develop features or representations from the raw text material appropriate for a wide range of NLP tasks.

Recently, neural network based deep learning methods have been shown to perform well on various NLP tasks such as language modeling, machine translation, part-of-speech tagging, named entity recognition, sentiment analysis, and paraphrase detection. The most attractive aspect of deep learning methods is their ability to perform these tasks without external hand-designed resources or time-intensive feature engineering. To this end, deep learning develops and makes use an important concept called "embedding," which refers to the representation of symbolic information in natural language text at word-level, phrase-level, and even sentence-level in terms of continuous-valued vectors.

The early work highlighting the importance of word embedding came from [62], [367], and [63], although the original form came from [26] as a side product of language modeling. Raw symbolic word representations are transformed from the sparse vectors via 1-of-V coding with a very high dimension (i.e., the vocabulary size V or its square or even its cubic) into low-dimensional, real-valued vectors via a neural network and then used for processing by subsequent neural network layers. The key advantage of using the continuous space to represent words (or phrases) is its distributed nature, which enables sharing or grouping the representations of words with a similar meaning. Such sharing is not possible in the original symbolic space, constructed by 1-of-V coding with a very high dimension, for representing words. Unsupervised learning is used where "context" of the word is used as the learning signal in neural networks. Excellent tutorials were recently given by Socher et al. [338, 340] to explain how the neural network is trained to perform word embedding. More recent work proposes new ways of learning word embeddings that better capture the semantics of words by incorporating both local and global document contexts and better account for homonymy and polysemy by learning multiple embeddings per word [169]. Also, there is strong evidence that the use of RNNs can also provide empirically good performance in learning word embeddings [245]. While the use of NNLMs, whose aim is to predict the future words in context, also induces word embeddings as its by-product, much simpler ways of achieving the embeddings are possible without the need to do word prediction. As shown by Collobert and Weston [62], the neural networks used for creating word embeddings need much smaller output units than the huge size typically required for NNLMs.

In the same early paper on word embedding, Collobert and Weston [62] developed and employed a convolutional network as the common model to simultaneously solve a number of classic problems including part-of-speech tagging, chunking, named entity tagging, semantic role identification, and similar word identification. More recent work reported in [61] further developed a fast, purely discriminative approach for parsing based on the deep recurrent convolutional architecture. Collobert et al. [63] provide a comprehensive review on ways of applying unified neural network architectures and related deep learning algorithms to solve NLP problems from "scratch," meaning that no traditional NLP methods are used to extract features. The theme of this line of work is to avoid task-specific, "man-made" feature engineering while providing versatility and unified features constructed automatically from deep learning applicable to all natural language processing tasks. The systems described in [63] automatically learn internal representations or word embedding from vast amounts of mostly unlabeled training data while performing a wide range of NLP tasks.

The recent work by Mikolov et al. [246] derives word embeddings by simplifying the NNLM described in Section 8.1. It is found that the NNLM can be successfully trained in two steps. First, continuous word vectors are learned using a simple model which eliminates the

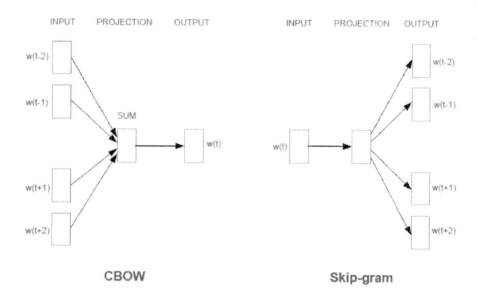

Figure 8.3: The CBOW architecture (a) on the left, and the Skip-gram architecture (b) on the right. [after [246], @ICLR].

nonlinearity in the upper neural network layer and share the projection layer for all words. And second, the N-gram NNLM is trained on top of the word vectors. So, after removing the second step in the NNLM, the simple model is used to learn word embeddings, where the simplicity allows the use of very large amount of data. This gives rise to a word embedding model called Continuous Bag-of-Words Model (CBOW), as shown in Figure 8.3a. Further, since the goal is no longer computing probabilities of word sequences as in LMs, the word embedding system here is made more effective by not only to predict the current word based on the context but also to perform inverse prediction known as "Skip-gram" model, as shown in Figure 8.3b. In the follow-up work [250] by the same authors, this word embedding system including the Skip-gram model is extended by a much faster learning method called negative sampling, similar to NCE discussed in Section 8.1.

In parallel with the above development, Mnih and Kavukcuoglu [254] demonstrate that NCE training of lightweight word embedding models is a highly efficient way of learning high-quality word representations, much like the somewhat earlier lightweight LMs developed by Mnih and Teh [255] described in Section 8.1. Consequently, results that used to require very considerable hardware and software infrastructure can now be obtained on a single desktop with minimal programming effort and using less time and data. This most recent work also shows that for representation learning, only five noise samples in NCE can be sufficient for obtaining strong results for word embedding, much fewer than that required for LMs. The authors also used an "inversed language model" for computing word embeddings, similar to the way in which the Skip-gram model is used in [250].

Huang et al. [169] recognized the limitation of the earlier work on word embeddings in that these models were built with only local context and one representation per word. They extended the local context models to one that can incorporate global context from full sentences or the entire document. This extended models accounts for homonymy and polysemy by learning multiple embeddings for each word. An illustration of this model is shown in Figure 8.4. In the earlier work by the same research group [344], a recursive neural network with local context was developed to build a deep architecture. The network, despite missing global context, was already shown to be capable of successful

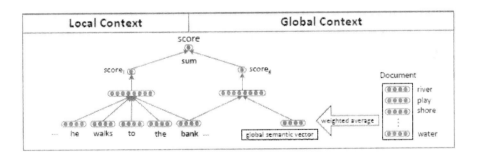

Figure 8.4: The extended word-embedding model using a recursive neural network that takes into account not only local context but also global context. The global context is extracted from the document and put in the form of a global semantic vector, as part of the input into the original word-embedding model with local context. Taken from Figure 1 of [169]. [after [169], @ACL].

merging of natural language words based on the learned semantic transformations of their original features. This deep learning approach provided an excellent performance on natural language parsing. The same approach was also demonstrated to be reasonably successful in parsing natural scene images. In related studies, a similar recursive deep architecture is used for paraphrase detection [346], and for predicting sentiment distributions from text [345].

We now turn to selected applications of deep learning methods including the use of neural network architectures and word embeddings to practically useful NLP tasks. Machine translation is one of such tasks, pursued by NLP researchers for many years based typically on shallow statistical models. The work described in [320] are perhaps the first comprehensive report on the successful application of neural-network-based language models with word embeddings, trained on a GPU, for large machine translation tasks. They address the problem of high computation complexity, and provide a solution that allows training 500 million words with 20 hours. Strong results are reported, with perplexity down from 71 to 60 in LMs and the corresponding BLEU score gained by 1.8 points using the neural-network-based language models with word embeddings compared with the best back-off LM.

A more recent study on applying deep learning methods to machine translation appears in [121, 123], where the phrase-translation component, rather than the LM component in the machine translation system is replaced by the neural network models with semantic word embeddings. As shown in Figure 8.5 for the architecture of this approach, a pair of source (denoted by f) and target (denoted by e) phrases are projected into continuous-valued vector representations in a low-dimensional latent semantic space (denoted by the two y vectors). Then their translation score is computed by the distance between the pair in this new space. The projection is performed by two deep neural networks (not shown here) whose weights are learned on parallel training data. The learning is aimed to directly optimize the quality of end-to-end machine translation results. Experimental evaluation has been performed on two standard Europarl translation tasks used by the NLP community, English–French and German–English. The results show that the new semantic-based phrase translation model significantly

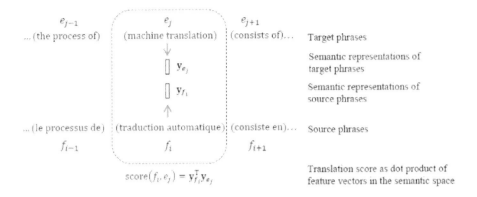

Figure 8.5: Illustration of the basic approach reported in [122] for machine translation. Parallel pairs of source (denoted by f) and target (denoted by e) phrases
are projected into continuous-valued vector representations (denoted by the two y
vectors), and their translation score is computed by the distance between the pair in
this continuous space. The projection is performed by deep neural networks (denoted
by the two arrows) whose weights are learned on parallel training data. [after [121],
@NIPS].

improves the performance of a state-of-the-art phrase-based statistical machine translation system, leading to a gain close to 1.0 BLEU
point.

A related approach to machine translation was developed by
Schwenk [320]. The estimation of the translation model probabilities of
a phrase-based machine translation system is carried out using neural
networks. The translation probability of phrase pairs is learned using
continuous-space representations induced by neural networks. A simplification is made that decomposes the translation probability of a
phrase or a sentence to a product of n-gram probabilities as in a standard n-gram language model. No joint representations of a phrase in
the source language and the translated version in the target language
are exploited as in the approach reported by Gao et al. [122, 123].

Yet another deep learning approach to machine translation
appeared in [249]. As in other approaches, a corpus of words in one
language are compared with the same corpus of words translated into
another, and words and phrases in such bilingual data that share similar
statistical properties are considered equivalent. A new technique is

proposed that automatically generates dictionaries and phrase tables that convert one language into another. It does not rely on versions of the same document in different languages. Instead, it uses data mining techniques to model the structure of a source language and then compares it to the structure of the target language. The technique is shown to translate missing word and phrase entries by learning language structures based on large monolingual data and mapping between languages from small bilingual data. It is based on vector-valued word embeddings as discussed earlier in this chapter and it learns a linear mapping between vector spaces of source and target languages.

An earlier study on applying deep learning techniques with DBNs was provided in [111] to attack a machine transliteration problem, a much easier task than machine translation. This type of deep architectures and learning may be generalized to the more difficult machine translation problem but no follow-up work has been reported. As another early NLP application, Sarikaya et al. [318] applied DNNs (called DBNs in the paper) to perform a natural language call–routing task. The DNNs use unsupervised learning to discover multiple layers of features that are then used to optimize discrimination. Unsupervised feature discovery is found to make DBNs far less prone to overfitting than the neural networks initialized with random weights. Unsupervised learning also makes it easier to train neural networks with many hidden layers. DBNs are found to produce better classification results than several other widely used learning techniques, e.g., maximum entropy and boosting based classifiers.

One most interesting NLP task recently tackled by deep learning methods is that of knowledge base (ontology) completion, which is instrumental in question-answering and many other NLP applications. An early work in this space came from [37], where a process is introduced to automatically learn structured distributed embeddings of knowledge bases. The proposed representations in the continuous-valued vector space are compact and can be efficiently learned from large-scale data of entities and relations. A specialized neural network architecture, a generalization of "Siamese" network, is used. In the follow-up work that focuses on multi-relational data [36], the semantic matching energy model is proposed to learn vector representations for both entities and relations. More recent work [340] adopts an alternative approach, based on the use of neural tensor networks, to attack the problem of reasoning over a large joint knowledge graph for relation classification. The knowledge graph is represented as triples of a relation between two entities, and the authors aim to develop a neural network model suitable for inference over such relationships. The model they presented is a neural tensor network, with one layer only. The network is used to represent entities in a fixed-dimensional vectors, which are created separately by averaging pre-trained word embedding vectors. It then learn the tensor with the newly added relationship element that describes the interactions among all the latent components in each of the relationships. The neural tensor network can be visualized in Figure 8.6, where each dashed box denotes one of the two slices of the tensor. Experimentally, the paper [340] shows that this tensor model can effectively classify unseen relationships in WordNet and FreeBase.

As the final example of deep learning applied successfully to NLP, we discuss here sentiment analysis applications based on recursive deep

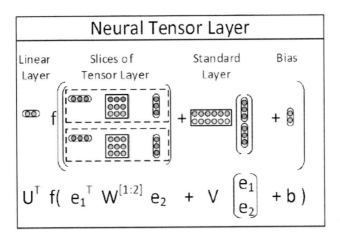

Figure 8.6: Illustration of the neural tensor network described in [340], with two relationships shown as two slices in the tensor. The tensor is denoted by $W^{[1:2]}$. The network contains a bilinear tensor layer that directly relates the two entity vectors (shown as e_1 and e_2) across three dimensions. Each dashed box denotes one of the two slices of the tensor. [after [340], @NIPS].

models published recently by Socher et al. [347]. Sentiment analysis is a task that is aimed to estimate the positive or negative opinion by an algorithm based on input text information. As we discussed earlier in this chapter, word embeddings in the semantic space achieved by neural network models have been very useful but it is difficult for them to express the meaning of longer phrases in a principled way. For sentiment analysis with the input data from typically many words and phrases, the embedding model requires the compositionality properties. To this end, Socher et al. [347] developed the recursive neural tensor network, where each layer is constructed similarly to that of the neural tensor network described in [340] with an illustration shown in Figure 8.6. The recursive construction of the full network exhibiting properties of compositionality follows that of [344] for the regular, non-tensor network. When trained on a carefully constructed sentiment analysis database, the recursive neural tensor network is shown to outperform all previous methods on several metrics. The new model pushes the state of the art in single sentence positive/negative classification accuracy from 80% up to 85.4%. The accuracy of predicting fine-grained sentiment labels for all phrases reaches 80.7%, an improvement of 9.7% over bag-of-features baselines.

9

Selected Applications in Information Retrieval

9.1 A brief introduction to information retrieval

Information retrieval (IR) is a process whereby a user enters a query into the automated computer system that contains a collection of many documents with the goal of obtaining a set of most relevant documents. Queries are formal statements of information needs, such as search strings in web search engines. In IR, a query does not uniquely identify a single document in the collection. Instead, several documents may match the query with different degrees of relevancy.

A document, sometimes called an object as a more general term which may include not only a text document but also an image, audio (music or speech), or video, is an entity that contains information and represented as an entry in a database. In this section, we limit the "object" to only text documents. User queries in IR are matched against the documents' representation stored in the database. Documents themselves often are not kept or stored directly in the IR system. Rather, they are represented in the system by metadata. Typical IR systems compute a numeric score on how well each document in the database matches the query, and rank the objects according to this value. The top-ranking documents from the system are then shown to the user. The process may then be iterated if the user wishes to refine the query.

Based partly on [236], common IR methods consist of several categories:

- Boolean retrieval, where a document either matches a query or does not.

- Algebraic approaches to retrieval, where models are used to represent documents and queries as vectors, matrices, or tuples. The similarity of the query vector and document vector is represented as a scalar value. This value can be used to produce a list of documents that are rank-ordered for a query. Common models and methods include vector space model, topic-based vector space model, extended Boolean model, and latent semantic analysis.

- Probabilistic approaches to retrieval, where the process of IR is treated as a probabilistic inference. Similarities are computed as probabilities that a document is relevant for a given query, and the probability value is then used as the score in ranking documents. Common models and methods include binary independence model, probabilistic relevance model with the BM25 relevance function, methods of inference with uncertainty, probabilistic, language modeling, http://en.wikipedia.org/wiki/Uncertain_inference and the technique of latent Dirichlet allocation.

- Feature-based approaches to retrieval, where documents are viewed as vectors of values of feature functions. Principled methods of "learning to rank" are devised to combine these features into a single relevance score. Feature functions are arbitrary functions of document and query, and as such Feature-based approaches can easily incorporate almost any other retrieval model as just yet another feature.

Deep learning applications to IR are rather recent. The approaches in the literature so far belong mostly to the category of feature-based approaches. The use of deep networks is mainly for extracting semantically meaningful features for subsequent document ranking stages.

We will review selected studies in the recent literature in the remainder of this section below.

9.2 Semantic hashing with deep autoencoders for document indexing and retrieval

Here we discuss the "semantic hashing" approach for the application of deep autoencoders to document indexing and retrieval as published in [159, 314]. It is shown that the hidden variables in the final layer of a DBN not only are easy to infer after using an approximation based on feed-forward propagation, but they also give a better representation of each document, based on the word-count features, than the widely used latent semantic analysis and the traditional TF-IDF approach for information retrieval. Using the compact code produced by deep autoencoders, documents are mapped to memory addresses in such a way that semantically similar text documents are located at nearby addresses to facilitate rapid document retrieval. The mapping from a word-count vector to its compact code is highly efficient, requiring only a matrix multiplication and a subsequent sigmoid function evaluation for each hidden layer in the encoder part of the network.

A deep generative model of DBN is exploited for the above purpose as discussed in [165]. Briefly, the lowest layer of the DBN represents the word-count vector of a document and the top layer represents a learned binary code for that document. The top two layers of the DBN form an undirected associative memory and the remaining layers form a Bayesian (also called belief) network with directed, top-down connections. This DBN, composed of a set of stacked RBMs as we reviewed in Section 5, produces a feed-forward "encoder" network that converts word-count vectors to compact codes. By composing the RBMs in the opposite order, a "decoder" network is constructed that maps compact code vectors into reconstructed word-count vectors. Combining the encoder and decoder, one obtains a deep autoencoder (subject to further fine-tuning as discussed in Section 4) for document coding and subsequent retrieval.

After the deep model is trained, the retrieval process starts with mapping each query into a 128-bit binary code by performing a forward pass through the model with thresholding. Then the Hamming distance between the query binary code and all the documents' 128-bit binary codes, especially those of the "neighboring" documents defined in the semantic space, are computed extremely efficiently. The efficiency is accomplished by looking up the neighboring bit vectors in the hash table. The same idea as discussed here for coding text documents for information retrieval has been explored for audio document retrieval and speech feature coding problems with some initial exploration reported in [100], discussed in Section 4 in detail.

9.3 Deep-structured semantic modeling (DSSM) for document retrieval

Here we discuss the more advanced and recent approach to large-scale document retrieval (Web search) based on a specialized deep architecture, called deep-structured semantic model or deep semantic similarity model (DSSM), as published in [172], and its convolutional version (C-DSSM), as published in [328].

Modern search engines retrieve Web documents mainly by matching keywords in documents with those in a search query. However, lexical matching can be inaccurate due to the fact that a concept is often expressed using different vocabularies and language styles in documents and queries. Latent semantic models are able to map a query to its relevant documents at the semantic level where lexical-matching often fails [236]. These models address the language discrepancy between Web documents and search queries by grouping different terms that occur in a similar context into the same semantic cluster. Thus, a query and a document, represented as two vectors in the lower-dimensional semantic space, can still have a high similarity even if they do not share any term. Probabilistic topic models such as probabilistic latent semantic models and latent Dirichlet allocation models have been proposed for semantic matching to partially overcome such difficulties. However, the improvement on IR tasks has not been as significant as originally expected because of two main factors: (1) most state-of-the-art latent semantic models are based on linear projection, and thus are inadequate in capturing effectively the complex semantic properties of documents;

and (2) these models are often trained in an unsupervised manner using an objective function that is only loosely coupled with the evaluation metric for the retrieval task. In order to improve semantic matching for IR, two lines of research have been conducted to extend the above latent semantic models. The first is the semantic hashing approach reviewed in Section 9.1 above in this section based on the use of deep autoencoders [165, 314]. While the hierarchical semantic structure embedded in the query and the document can be extracted via deep learning, the deep learning approach used for their models still adopts an unsupervised learning method where the model parameters are optimized for the re-construction of the documents rather than for differentiating the relevant documents from the irrelevant ones for a given query. As a result, the deep neural network models do not significantly outperform strong baseline IR models that are based on lexical matching. In the second line of research, click-through data, which consists of a list of queries and the corresponding clicked documents, is exploited for semantic modeling so as to bridge the language discrepancy between search queries and Web documents in recent studies [120, 124]. These models are trained on click-through data using objectives that tailor to the document ranking task. However, these click-through-based models are still linear, suffering from the issue of expressiveness. As a result, these models need to be combined with the keyword matching models (such as BM25) in order to obtain a significantly better performance than baselines.

The DSSM approach reported in [172] aims to combine the strengths of the above two lines of work while overcoming their weaknesses. It uses the DNN architecture to capture complex semantic properties of the query and the document, and to rank a set of documents for a given query. Briefly, a nonlinear projection is performed first to map the query and the documents to a common semantic space. Then, the relevance of each document given the query is calculated as the cosine similarity between their vectors in that semantic space. The DNNs are trained using the click-through data such that the conditional likelihood of the clicked document given the query is maximized. Different from the previous latent semantic models that are learned in an unsupervised fashion, the DSSM is optimized directly for Web

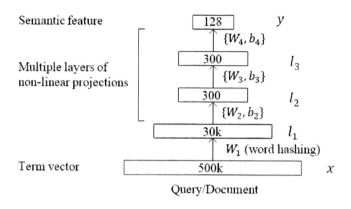

Figure 9.1: The DNN component of the DSSM architecture for computing semantic features. The DNN uses multiple layers to map high-dimensional sparse text features, for both Queries and Documents into low-dimensional dense features in a semantic space. [after [172], @CIKM].

document ranking, and thus gives superior performance. Furthermore, to deal with large vocabularies in Web search applications, a new *word hashing* method is developed, through which the high-dimensional term vectors of queries or documents are projected to low-dimensional letter based *n*-gram vectors with little information loss.

Figure 9.1 illustrates the DNN part in the DSSM architecture. The DNN is used to map high-dimensional sparse text features into low-dimensional dense features in a semantic space. The first hidden layer, with 30k units, accomplishes word hashing. The word-hashed features are then projected through multiple layers of non-linear projections. The final layer's neural activities in this DNN form the feature in the semantic space.

To show the computational steps in the various layers of the DNN in Figure 9.1, we denote x as the input term vector, y as the output vector, l_i, $i = 1, \ldots, N - 1$, as the intermediate hidden layers, W_i as the ith projection matrix, and b_i as the ith bias vector, we have

$$
\begin{aligned}
l_1 &= W_1 x, \\
l_i &= f(W_i l_{i-1} + b_i), \quad i > 1 \\
y &= f(W_N l_{N-1} + b_N),
\end{aligned}
$$

where $tanh$ function is used at the output layer and the hidden layers $l_i, i = 2, \ldots, N - 1$:

$$f(x) = \frac{1 - e^{-2x}}{1 + e^{-2x}}.$$

The semantic relevance score between a query Q and a document D can then be computed as the consine distance

$$R(Q, D) = \text{cosine}(y_Q, y_D) = \frac{y_Q^T y_D}{\|y_Q\| \|y_D\|},$$

where y_Q and y_D are the concept vectors of the query and the document, respectively. In Web search, given the query, the documents can be sorted by their semantic relevance scores.

Learning of the DNN weights W_i and b_i shown in Figure 9.1 is an important contribution of the study of [172]. Compared with the DNNs used in speech recognition where the targets or labels of the training data are readily available, the DNN in the DSSM does not have such label information well defined. That is, rather than using the common cross entropy or mean square errors as the training objective function, IR-centric loss functions need to be developed in order to train the DNN weights in the DSSM using the available data such as click-through logs.

The click-through logs consist of a list of queries and their clicked documents. A query is typically more relevant to the documents that are clicked on than those that are not. This weak supervision information can be exploited to train the DSSM. More specifically, the weight matrices in the DSSM, W_i, is learned to maximize the posterior probability of the clicked documents given the queries

$$P(D|Q) = \frac{\exp(\gamma R(Q, D))}{\sum_{D' \in \mathbf{D}} \exp(\gamma R(Q, D'))}$$

defined on the semantic relevance score $R(Q, D)$ between the Query (Q) and the Document (D), where γ is a smoothing factor set empirically on a held-out data set, and $\mathbf{D}$ denotes the set of candidate documents to be ranked. Ideally, $\mathbf{D}$ should contain all possible documents, as in the maximum mutual information training for speech recognition where all possible negative candidates may be considered [147]. However in this case **D** is of Web scale and thus is intractable in practice. In the implementation of DSSM learning described in [172], a subset of the negative candidates are used, following the common practice adopted in MCE (Minimum Classification Error) training in speech recognition [52, 118, 417, 418]. In other words, for each query and clicked-document pair, denoted by (QD^+) where Q is a query and D^+ is the clicked document, the set of **D** is approximated by including D^+ and only four randomly selected unclicked documents, denoted by $D_j^-; j = 1, \ldots, 4\}$. In the study reported in [172], no significant difference was found when different sampling strategies were used to select the unclicked documents.

With the above simplification the DSSM parameters are estimated to maximize the approximate likelihood of the clicked documents given the queries across the training set

$$L(\Lambda) = \log \prod_{(Q,D^+,D_j^-)} P(D^+ \mid Q),$$

where Λ denotes the parameter set of the DNN weights $\{W_i\}$ in the DSSM. In Figure 9.2, we show the overall DSSM architecture that contains several DNNs. All these DNNs share the same weights but take different documents (one positive and several negatives) as inputs when training the DSSM parameters. Details of the gradient computation of this approximate loss function with respect to the DNN weights tied across documents and queries can be found in [172] and are not elaborated here.

Most recently, the DSSM described above has been extended to its convolutional version, or C-DSSM [328]. In the C-DSSM, semantically similar words within context are projected to vectors that are close to each other in the contextual feature space through a convolutional structure. The overall semantic meaning of a sentence is found to be determined by a few *key* words in the sentence, and thus the C-DSSM uses an additional max pooling layer to extract the most salient local features to form a fixed-length global feature vector. The global feature vector is then fed to the remaining nonlinear DNN layer(s) to map it to a point in the shared semantic space.

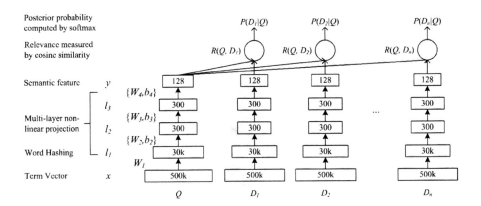

Figure 9.2: Architectural illustration of the DSSM for document retrieval (from [170, 171]). All DNNs shown have shared weights. A set of n documents are shown here to illustrate the random negative sampling discussed in the text for simplifying the training procedure for the DSSM. [after [172], @CIKM].

The convolutional neural network component of the C-DSSM is shown in Figure 9.3, where a window size of three is illustrated for the convolutional layer. The overall C-DSSM architecture is similar to the DSSM architecture shown in Figure 9.2 except that the fully-connected DNNs are replaced by the convolutional neural networks with locally-connected tied weights and additional max-pooling layers. The model component shown in Figure 9.3 contains (1) a word hashing layer to transform words into letter-tri-gram count vectors in the same way as the DSSM; (2) a convolutional layer to extract local contextual features for each context window; (3) a max-pooling layer to extract and combine salient local contextual features to form a global feature vector; and (4) a semantic layer to represent the high-level semantic information of the input word sequence.

The main motivation for using the convolutional structure in the C-DSSM is its ability to map a variable-length word sequence to a low-dimensional vector in a latent semantic space. Unlike most previous models that treat a query or a document as a bag of words, a query or a document in the C-DSSM is viewed as a sequence of words with contextual structures. By using the convolutional structure, local contextual information at the word n-gram level is modeled first. Then,

Semantic layer: y

Affine projection matrix: W_s

Max pooling layer: v

Max pooling operation

Convolutional layer: h_t

Convolution matrix: W_c

Word hashing layer: f_t

Word hashing matrix: W_f

Word sequence: x_t

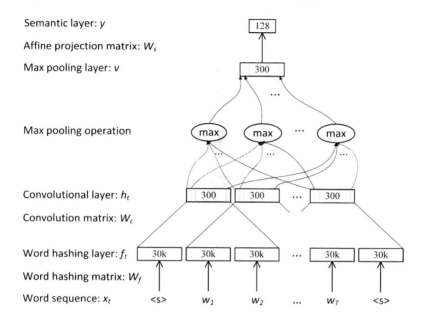

Figure 9.3: The convolutional neural network component of the C-DSSM, with the window size of three is illustrated for the convolutional layer. [after [328], @WWW].

salient local features in a word sequence are combined to form a global feature vector. Finally, the high-level semantic information of the word sequence is extracted to form a global vector representation. Like the DSSM just described, the C-DSSM is also trained on click-through data by maximizing the conditional likelihood of the clicked documents given a query using the back-propagation algorithm.

9.4   Use of deep stacking networks for information retrieval

In parallel with the IR studies reviewed above, the deep stacking network (DSN) discussed in Section 6 has also been explored recently for IR with insightful results [88]. The experimental results suggest that the classification error rate using the binary decision of "relevant" versus "non-relevant" from the DSN, which is closely correlated with the DSN training objective, is also generally correlated well with the NDCG (normalized discounted cumulative gain) as the most common

IR quality measure. The exception is found in the region of high IR quality.

As described in Section 6, the simplicity of the DSN's training objective, the mean square error (MSE), drastically facilitates its successful applications to image recognition, speech recognition, and speech understanding. The MSE objective and classification error rate have been shown to be well correlated in these speech or image applications. For information retrieval (IR) applications, however, the inconsistency between the MSE objective and the desired objective (e.g., NDCG) is much greater than that for the above classification-focused applications. For example, the NDCG as a desirable IR objective function is a highly non-smooth function of the parameters to be learned, with a very different nature from the nonlinear relationship between MSE and classification error rate. Thus, it is of interest to understand to what extent the NDCG is reasonably well correlated with classification rate or MSE where the relevance level in IR is used as the DSN prediction target. Further, can the advantage of learning simplicity in the DSN be applied to improve IR quality measures such as the NDCG? Our experimental results presented in [88] provide largely positive answers to both of the above questions. In addition, special care that need to be taken in implementing DSN learning algorithms when moving from classification to IR applications are addressed.

The IR task in the experiments of [88] is the sponsored search related to ad placement. In addition to the organic web search results, commercial search engines also provide supplementary sponsored results in response to the user's query. The sponsored search results are selected from a database pooled by advertisers who bid to have their ads displayed on the search result pages. Given an input query, the search engine will retrieve relevant ads from the database, rank them, and display them at the proper place on the search result page; e.g., at the top or right hand side of the web search results. Finding relevant ads to a query is quite similar to common web search. For instance, although the documents come from a constrained database, the task resembles typical search ranking that targets on predicting document relevance to the input query. The experiments conducted for this task are the first with the use of deep learning techniques (based on the DSN architecture) on the ad-related IR problem. The preliminary results from the experiments are the close correlation between the MSE as the DSN training objective with the NDCG as the IR quality measure over a wide NDCG range.

10

Selected Applications in Object Recognition and Computer Vision

Over the past two years or so, tremendous progress has been made in applying deep learning techniques to computer vision, especially in the field of object recognition. The success of deep learning in this area is now commonly accepted by the computer vision community. It is the second area in which the application of deep learning techniques is successful, following the speech recognition area as we reviewed and analyzed in Sections 2 and 7.

Excellent surveys on the recent progress of deep learning for computer vision are available in the NIPS-2013 tutorial (https://nips.cc/Conferences/2013/Program/event.php?ID=4170 with video recording at http://research.microsoft.com/apps/video/default.aspx?id=206976&l=i) and slides at http://cs.nyu.edu/~fergus/presentations/nips2013_final.pdf, and also in the CVPR-2012 tutorial (http://cs.nyu.edu/~fergus/tutorials/deep_learning_cvpr12). The reviews provided in this section below are based partly on these tutorials, in connection with the earlier deep learning material in this monograph. Another excellent source which this section draws from is the most recent Ph.D. thesis on the topic of deep learning for computer vision [434].

Over many years, object recognition in computer vision has been relying on hand-designed features such as SIFT (scale invariant feature transform) and HOG (histogram of oriented gradients), akin to the reliance of speech recognition on hand-designed features such as MFCC and PLP. However, features like SIFT and HOG only capture low-level edge information. The design of features to effectively capture mid-level information such as edge intersections or high-level representation such as object parts becomes much more difficult. Deep learning aims to overcome such challenges by automatically learning hierarchies of visual features in both unsupervised and supervised manners directly from data. The review below categorizes the many deep learning methods applied to computer vision into two classes: (1) unsupervised feature learning where the deep learning is used to extract features only, which may be subsequently fed to relatively simple machine learning algorithm for classification or other tasks; and (2) supervised learning methods where end-to-end learning is adopted to jointly optimize feature extractor and classifier components of the full system when large amounts of labeled training data are available.

10.1   Unsupervised or generative feature learning

When labeled data are relatively scarce, unsupervised learning algorithms have been shown to learn useful visual feature hierarchies. In fact, prior to the demonstration of remarkable successes of CNN architectures with supervised learning in the 2012 ImageNet competition, much of the work in applying deep learning methods to computer vision had been on unsupervised feature learning. The original unsupervised deep autoencoder that exploits DBN pre-training was developed and demonstrated by Hinton and Salakhutdinov [164] with success on the image recognition and dimensionality reduction (coding) tasks of MNIST with only 60,000 samples in the training set; see details of this task in http://yann.lecun.com/exdb/mnist/ and an analysis in [78]. It is interesting to note that the gain of coding efficiency using the DBN-based autoencoder on the image data over the conventional method of principal component analysis as demonstrated in [164] is very similar to the gain reported in [100] and described in Section 4 of this monograph on the speech data over the traditional technique of vector quantization. Also, Nair and Hinton [265] developed a modified DBN where the top-layer model uses a third-order Boltzmann machine. This type of DBN is applied to the NORB database — a three-dimensional object recognition task. An error rate close to the best published result on this task is reported. In particular, it is shown that the DBN substantially outperforms shallow models such as SVMs. In [358], two strategies to improve the robustness of the DBN are developed. First, sparse connections in the first layer of the DBN are used as a way to regularize the model. Second, a probabilistic de-noising algorithm is developed. Both techniques are shown to be effective in improving robustness against occlusion and random noise in a noisy image recognition task. DBNs have also been successfully applied to create compact but meaningful representations of images [360] for retrieval purposes. On this large collection image retrieval task, deep learning approaches also produced strong results. Further, the use of a temporally conditional DBN for video sequence and human motion synthesis were reported in [361]. The conditional RBM and DBN make the RBM and DBN weights associated with a fixed time window conditioned on the data from previous time steps. The computational tool offered in this type of temporal DBN and the related recurrent networks may provide the opportunity to improve the DBN–HMMs towards efficient integration of temporal-centric human speech production mechanisms into DBN-based speech production model.

Deep learning methods have a rich family, including hierarchical probabilistic and generative models (neural networks or otherwise). One most recent example of this type developed and applied to facial expression datasets is the stochastic feed-forward neural networks that can be learned efficiently and that can induce a rich multiple-mode distribution in the output space not possible with the standard, deterministic neural networks [359]. In Figure 10.1, we show the architecture of a typical stochastic feed-forward neural network with four hidden layers with mixed deterministic and stochastic neurons (left) used to model multi-mode distributions illustrated on the right. The stochastic network here is a deep, directed graphical model, where the generation

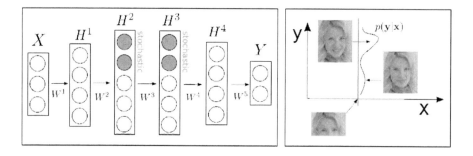

Figure 10.1: Left: A typical architecture of the stochastic feed-forward neural network with four hidden layers. Right: Illustration of how the network can produce a distribution with two distinct modes and use them to represent two or more different facial expressions **y** given a neutral face **x**. [after [359], @NIPS].

process starts from input **x**, a neural face, and generates the output **y**, the facial expression. In face expression classification experiments, the learned unsupervised hidden features generated from this stochastic network are appended to the image pixels and helped to obtain superior accuracy to the baseline classifier based on the conditional RBM/DBN [361].

Perhaps the most notable work in the category of unsupervised deep feature learning for computer vision (prior to the recent surge of the work on CNNs) is that of [209], a nine-layer locally connected sparse autoencoder with pooling and local contrast normalization. The model has one billion connections, trained on the dataset with 10 million images downloaded from the Internet. The unsupervised feature learning methods allow the system to train a face detector without having to label images as containing a face or not. And the control experiments show that this feature detector is robust not only to translation but also to scaling and out-of-plane rotation.

Another set of popular studies on unsupervised deep feature learning for computer vision are based on deep sparse coding models [226]. This type of deep models produced state-of-the-art accuracy results on the ImageNet object recognition tasks prior to the rise of the CNN architectures armed with supervised learning to perform joint feature learning and classification, which we turn to now.

10.2   Supervised feature learning and classification

The origin of the applications of deep learning to object recognition tasks can be traced to the convolutional neural networks (CNNs) in the early 90s; see a comprehensive overview in [212]. The CNN-based architectures in the supervised learning mode have captured intense interest in computer vision since October 2012 shortly after the ImageNet competition results were released (http://www.image-net.org/challenges/LSVRC/2012/). This is mainly due to the huge recognition accuracy gain over competing approaches when large amounts of labeled data are available to efficiently train large CNNs using GPU-like high-performance computing platforms. Just like DNN-based deep learning methods have outperformed previous state-of-the-art approaches in speech recognition in a series of benchmark tasks including phone recognition, large-vocabulary speech recognition, noise-robust speech recognition, and multi-lingual speech recognition, CNN-based deep learning methods have demonstrated the same in a set of computer vision benchmark tasks including category-level object recognition, object detection, and semantic segmentation.

The basic architecture of the CNN described in [212] is shown in Figure 10.1. To incorporate the relative invariance of the spatial relationship in typical image pixels with respect to the location, the CNN uses a convolutional layer with local receptive fields and with tied filter weights, much like 2-dimensional FIR filters in image processing. The output of the FIR filters is then passed through a nonlinear activation function to create activation maps, followed by another nonlinear pooling (labeled as "subsampling" in Figure 10.2) layer that reduces the data rate while providing invariance to slightly different input images. The output of the pooling layer is fed to a few fully connected layers as in the DNN discussed in earlier chapters. The whole architecture above is also called the deep CNN in the literature.

Deep models with convolution structure such as CNNs have been found effective and have been in use in computer vision and image recognition since 90s [57, 185, 192, 198, 212]. The most notable advance was achieved in the 2012 ImageNet LSVRC competition, in which

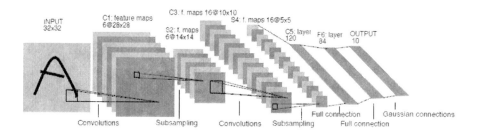

Figure 10.2: The original convolutional neural network that is composed of multiple alternating convolution and pooling layers followed by fully connected layers. [after [212], @IEEE].

the task is to train a model with 1.2 million high-resolution images to classify unseen images to one of the 1000 different image classes. On the test set consisting of 150k images, the deep CNN approach described in [198] achieved the error rates considerably lower than the previous state-of-the-art. Very large deep-CNNs are used, consisting of 60 million weights, and 650,000 neurons, and five convolutional layers together with max-pooling layers. Additional two fully-connected layers as in the DNN described previously are used on top of the CNN layers. Although all the above structures were developed separately in earlier work, their best combination accounted for major part of the success. See the overall architecture of the deep CNN system in Figure 10.3. Two additional factors contribute to the final success. The first is a powerful regularization technique called "dropout"; see details in [166] and a series of further analysis and improvement in [10, 13, 240, 381, 385]. In particular, Warde-Farley et al. [385] analyzed the disentangling effects of dropout and showed that it helps because different members of the bag share parameters. Applications of the same "dropout" techniques are also successful for some speech recognition tasks [65, 81]. The second factor is the use of non-saturating neurons or rectified linear units (ReLU) that compute $f(x) = \max(x, 0)$, which significantly speeds up the overall training process especially with efficient GPU implementation. This deep-CNN system achieved a winning top-5 test error rate of 15.3% using extra training data from ImageNet Fall 2011 release, or 16.4% using only supplied training data in ImageNet-2012,

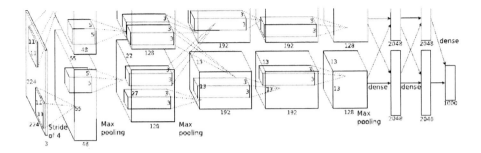

Figure 10.3: The architecture of the deep-CNN system which won the 2012 ImageNet competition by a large margin over the second-best system and the state of the art by 2012. [after [198], @NIPS].

significantly lower than 26.2% achieved by the second-best system which combines scores from many classifiers using a set of hand-crafted features such as SIFT and Fisher vectors. See details in http://www.image-net.org/challenges/LSVRC/2012/oxford_vgg.pdf about the best competing method. It is noted, however, that the Fisher-vector-encoding approach has recently been extended by Simonyan et al. [329] via stacking in multiple layers to form deep Fisher networks, which achieve competitive results with deep CNNs at a smaller computational learning cost.

The state of the art performance demonstrated in [198] using the deep-CNN approach is further improved by another significant margin during 2013, using a similar approach but with bigger models and larger amounts of training data. A summary of top-5 test error rates from 11 top-performing teams participating in the 2013 ImageNet ILSVRC competition is shown in Figure 10.4, with the best result of the 2012 competition shown to the right most as the baseline. Here we see rapid error reduction on the same task from the lowest pre-2012 error rate of 26.2% (non-neural networks) to 15.3% in 2012 and further to 11.2% in 2013, both achieved with deep-CNN technology. It is also interesting to observe that all major entries in the 2013 ImageNet ILSVRC competition is based on deep learning approaches. For example, the Adobe system shown in Figure 10.4 is based on the deep-CNN reported in [198] including the use of dropout. The network

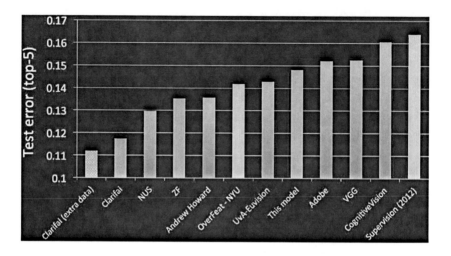

Figure 10.4: Summary results of ImageNet Large Scale Visual Recognition Challenge 2013 (ILSVRC2013), representing the state-of-the-are performance of object recognition systems. Data source: http://www.image-net.org/challenges/ LSVRC/2013/results.php.

architecture is modified to include more filters and connections. At test time, image saliency is used to obtain 9 crops from original images, which are combined with the standard five multiview crops. The NUS system uses a non-parametric, adaptive method to combine the outputs from multiple shallow and deep experts, including deep-CNN, kernel, and GMM methods. The VGG system is described in [329] and uses a combination of the deep Fisher vector network and the deep-CNN. The ZF system is based on a combination of a large CNN with a range of different architectures. The choice of architectures was assisted by visualization of model features using a deconvolutional network as described by Zeiler et al. [437], Zeiler and Fergus [435, 436], and Zeiler ([434]). The CognitiveVision system uses an image classification scheme based on a DNN architecture. The method is inspired by cognitive psychophysics about how the human vision system first learns to classify the basic-level categories and then learns to classify categories at the subordinate level for fine-grained object recognition. Finally, the best-performing system called Clarifai in Figure 10.4 is based on a large and deep CNN with dropout regularization. It augments the amount of training data by down-sampling 256 pixels. The system contains a total of 65M parameters. Multiple such models were averaged together to further boost performance. The main novelty is to use the visualization technique based on the deconvolutional networks as described in [434, 437] to identify what makes the deep model perform well, based on which a powerful deep architecture was chosen. See more details of these systems in http://www.image-net.org/challenges/LSVRC/2013/results.php.

While the deep CNN has demonstrated remarkable classification performance on object recognition tasks, there has been no clear understanding of why they perform so well until recently. Zeiler and Fergus [435, 436] conducted research to address just this issue, and then used the gained understanding to further improve the CNN systems, which yielded excellent performance as shown in Figure 10.4 with labels "ZF" and "Clarifai." A novel visualization technique is developed that gives insight into the function of intermediate feature layers of the deep CNN. The technique also sheds light onto the operation of the full network acting as a classifier. The visualization technique is based on a deconvolutional network, which maps the neural activities in intermediate layers of the original convolutional network back to the input pixel space. This allows the researchers to examine what input pattern originally caused a given activation in the feature maps. Figure 10.5 (the top portion) illustrates how a deconvolutional network is attached to each of its layers, thereby providing a closed loop back to image pixels as the input to the original CNN. The information flow in this closed loop is as follows. First, an input image is presented to the deep CNN in a feed-forward manner so that the features at all layers are computed. To examine a given CNN activation, all other activations in the layer are set to zero and the feature maps are passed as input to the attached deconvolutional network's layer. Then, successive operations, opposite to the feed-forward computation in the CNN, are carried out including unpooling, rectifying, and filtering. This allows the reconstruction of the activity in the layer beneath that gave rise to the chosen activation. These operations are repeated until input layer is reached. During unpooling, non-invertibility of the max pooling operation in the CNN is

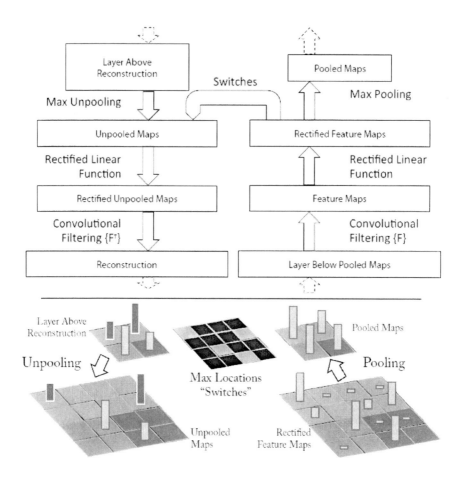

Figure 10.5: The top portion shows how a deconvolutional network's layer (left) is attached to a corresponding CNN's layer (right). The *d* econvolutional network reconstructs an approximate version of the CNN features from the layer below. The bottom portion is an illustration of the unpooling operation in the deconvolutional network, where "Switches" are used to record the location of the local max in each pooling region during pooling in the CNN. [after [436], @arXiv].

resolved by an approximate inverse, where the locations of the maxima within each pooling region are recorded in a set of "switch" variables. These switches are used to place the reconstructions from the layer above into appropriate locations, preserving the structure of the stimulus. This procedure is shown at the bottom portion of Figure 10.5.

In addition to the deep-CNN architecture described above, the DNN architecture has also been shown to be highly successful in a number of computer vision tasks [54, 55, 56, 57]. We have not found in the literature on direct comparisons among the CNN, DNN, and other related architectures on the identical tasks.

Finally, the most recent study on supervised learning for computer vision shows that the deep CNN architecture is not only successful for object/image classification discussed earlier in this section but also successful for objection detection in the whole images [128]. The detection task is substantially more complex than the classification task.

As a brief summary of this chapter, deep learning has made huge inroads into computer vision, soon after its success in speech recognition discussed in Section 7. So far, it is the supervised learning paradigm based on the deep CNN architecture and the related classification techniques that are making the greatest impact, showcased by the ImageNet competition results from 2012 and 2013. These methods can be used for not only object recognition but also many other computer vision tasks. There has been some debate as to the reasons for the success of these CNN-based deep learning methods, and about their limitations. Many questions are still open as to how these methods can be tailored to certain computer vision applications and how to scale up the models and training data. Finally, we discussed a number of studies on unsupervised and generative approaches of deep learning to computer vision and image modeling problems in the earlier part of this chapter. Their performance has not been competitive with the supervised learning approach on object recognition tasks with ample training data. To achieve long term and ultimate success in computer vision, it is likely that unsupervised learning will be needed. To this end, many open problems in unsupervised feature learning and deep learning need to be addressed and much more research need to be carried out.

11

Selected Applications in Multimodal and Multi-task Learning

Multi-task learning is a machine learning approach that learns to solve several related problems at the same time, using a shared representation. It can be regarded as one of the two major classes of transfer learning or learning with knowledge transfer, which focuses on generalizations across distributions, domains, or tasks. The other major class of transfer learning is adaptive learning, where knowledge transfer is carried out in a sequential manner, typically from a source task to a target task [95]. Multi-modal learning is a closely related concept to multi-task learning, where the learning domains or "tasks" cut across several modalities for human–computer interactions or other applications embracing a mixture of textual, audio/speech, touch, and visual information sources.

The essence of deep learning is to automate the process of discovering effective features or representations for any machine learning task, including automatically transferring knowledge from one task to another concurrently. Multi-task learning is often applied to conditions where no or very little training data are available for the target task domain, and hence is sometimes called zero-shot or one-shot learning. It is evident that difficult multi-task leaning naturally fits the paradigm of deep learning or representation learning where the shared

s and statistical strengths across tasks (e.g., those involv-
odalities of audio, image, touch, and text) is expected
greatly facilitate many machine learning scenarios under low- or
zero-resource conditions. Before deep learning methods were adopted,
there had been numerous efforts in multi-modal and multi-task learning. For example, a prototype called MiPad for multi-modal interactions involving capturing, leaning, coordinating, and rendering a mix
of speech, touch, and visual information was developed and reported
in [175, 103]. And in [354, 443], mixed sources of information from
multiple-sensory microphones with separate bone-conductive and air-
born paths were exploited to de-noise speech. These early studies all
used shallow models and learning methods and achieved worse than
desired performance. With the advent of deep learning, it is hopeful
that the difficult multi-modal learning problems can be solved with
eventual success to enable a wide range of practical applications. In
this chapter, we will review selected applications in this area, organized according to different combinations of more than one modalities
or learning tasks. Much of the work reviewed here is on-going research,
and readers should expect follow-up publications in the future.

11.1   Multi-modalities: Text and image

The underlying mechanism for potential effectiveness of multi-modal
learning involving text and image is the common semantics associated
with the text and image. The relationship between the text and image
may come, for example, from the text annotations of an image (as the
training data for a multi-modal learning system). If the related text
and image share the same representation in a common semantic space,
the system can generalize to the unseen situation where either text
or image is unavailable. It can thus be naturally used for zero-shot
learning for image or text. In other words, multi-modality learning can
use text information to help image/visual recognition, and vice versa.
Exploiting text information for image/visual recognition constitutes
most of the work done in this space, which we review in this section
below.

The deep architecture, called DeViSE (deep visual-semantic embedding) and developed by Frome et al. [117], is a typical example of the multi-modal learning where text information is used to improve the image recognition system, especially for performing zero-shot learning. Image recognition systems are often limited in their ability to scale to large number of object categories, due in part to the increasing difficulty of acquiring sufficient training data with text labels as the number of image categories grows. The multi-modal DeViSE system is aimed to leverage text data to train the image models. The joint model is trained to identify image classes using both labeled image data and the semantic information learned from unannotated text. An illustration of the DeViSE architecture is shown in the center portion of Figure 10.1. It is initialized with the parameters pre-trained at the lower layers of two models: the deep-CNN for image classification in the left portion of the figure and the text embedding model in the right portion of the figure. The part of the deep CNN, labeled "core visual model" in Figure 10.1, is further learned to predict the target word-embedding vector using a projection layer labeled "transformation" and using a similarity metric. The loss function used in training adopts a combination of dot-product similarity and max-margin, hinge rank loss. The former is the un-normalized version of the cosine loss function used for training the DSSM model in [170] as described in Section 9.3. The latter is similar to the earlier joint image-text model called WSABIE (web scale annotation by image embedding developed by Weston et al. [388, 389]. The results show that the information provided by text improves zero-shot image predictions, achieving good hit rates (close to 15%) across thousands of the labels never seen by the image model.

The earlier WSABIE system as described in [388, 389] adopted a shallow architecture and trained a joint embedding model of both images and labels. Rather than using deep architectures to derive the highly nonlinear image (as well as text-embedding) feature vectors as in DeViSE, the WSABIE uses simple image features and a linear mapping to arrive at the joint embedding space. Further, it uses an embedding vector for each possible label. Thus, unlike DeViSE, WSABIE could not generalize to new classes.

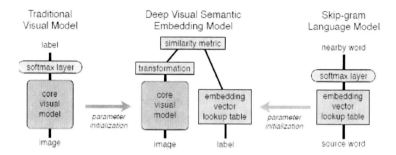

Figure 11.1: Illustration of the multi-modal DeViSE architecture. The left portion is an image recognition neural network with a softmax output layer. The right portion is a skip-gram text model providing word embedding vectors; see Section 8.2 and Figure 8.3 for details. The center is the joint deep image-text model of DeViSE, with the two Siamese branches initialized by the image and word embedding models below the softmax layers. The layer labeled "transformation" is responsible for mapping the outputs of the image (left) and text (right) branches into the same semantic space. [after [117], @NIPS].

It is also interesting to compare the DeViSE architecture of Figure 11.1 with the DSSM architecture of Figure 9.2 in Section 9. The branches of "Query" and "Documents" in DSSM are analogous to the branches of "image" and "text-label" in DeViSE. Both DeViSE and DSSM use the objective function related to cosine distance between two vectors for training the network weights in an end-to-end fashion. One key difference, however, is that the two sets of inputs to the DSSM are both text (i.e., "Query" and "Documents" designed for IR), and thus mapping "Query" and "Documents" to the same semantic space is conceptually more straightforward compared with the need in DeViSE for mapping from one modality (image) to another (text). Another key difference is that the generalization ability of DeViSE to unseen image classes comes from computing text embedding vectors for many unsupervised text sources (i.e., with no image counterparts) that would cover the text labels corresponding to the unseen classes. The generalization ability of the DSSM over unseen words, however, is derived from a special coding scheme for words in terms of their constituent letters.

The DeViSE architecture has inspired a more recent method, which maps images into the semantic embedding space via convex combination of embedding vectors for the text label and the image classes [270]. Here is the main difference. DeViSE replaces the last, softmax layer of a CNN image classifier with a linear transformation layer. The new transformation layer is then trained together with the lower layers of the CNN. The method in [270] is much simpler — keeping the softmax layer of the CNN while not training the CNN. For a test image, the CNN first produces top N-best candidates. Then, the convex combination of the corresponding N embedding vectors in the semantic space is computed. This gives a deterministic transformation from the outputs of the softmax classifier into the embedding space. This simple multi-modal learning method is shown to work very well on the ImageNet zero-shot learning task.

Another thread of studies separate from but related to the above work on multi-modal learning involving text and image have centered on the use of multi-modal embeddings, where data from multiple sources with separate modalities of text and image are projected into the same vector space. For example, Socher and Fei-Fei [341] project words and images into the same space using kernelized canonical correlation analysis. Socher et al. [342] map images to single-word vectors so that the constructed multi-modal system can classify images without seeing any examples of the class, i.e., zero-shot learning similar to the capability of DeViSE. The most recent work by Socher et al. [343] extends their earlier work from single-word embeddings to those of phrases and full-length sentences. The mechanism for mapping sentences instead of the earlier single words into the multi-modal embedding space is derived from the power of the recursive neural network described in Socher et al. [347] as summarized in Section 8.2, and its extension with dependency tree.

In addition to mapping text to image (or vice versa) into the same vector space or to creating the joint image/text embedding space, multi-modal learning for text and image can also be cast in the framework of language models. In [196], a model of natural language is made conditioned on other modalities such as image as the focus of the study. This type of multi-modal language model is used to (1) retrieve images given complex description queries, (2) retrieve phrase descriptions given image queries, and (3) generate text conditioned on images.

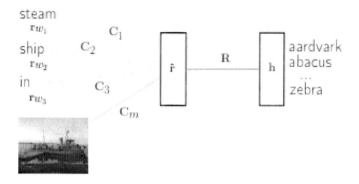

Figure 11.2: Illustration of the multi-modal DeViSE architecture. The left portion is an image recognition neural network with a softmax output layer. The right portion is a skip-gram text model providing word embedding vectors; see Section 8.2 and Figure 8.3 for details. The center is the joint deep image-text model of DeViSE, with the two Siamese branches initialized by the image and word embedding models below the softmax layers. The layer labeled "transformation" is responsible for mapping the outputs of the image (left) and text (right) branches into the same semantic space. [after [196], @NIPS].

Word representations and image features are jointly learned by train-ing the multi-modal language model together with a convolutional net-work. An illustration of the multi-modal language model is shown in Figure 11.2.

11.2   Multi-modalities: Speech and image

Ngiam et al. [268, 269] propose and evaluate an application of deep networks to learn features over audio/speech and image/video modalities. They demonstrate cross-modality feature learning, where better features for one modality (e.g., image) is learned when multiple modalities (e.g., speech and image) are present at feature learning time. A bi-modal deep autoencoder architecture for separate audio/speech and video/image input channels are shown in Figure 11.3. The essence of this architecture is to use a shared, middle layer to represent both types of modalities. This is a straightforward generalization from the single-modal deep autoencoder for speech shown in Figure 4.1 of Section 4 to bi-modal counterpart. The authors further show how to

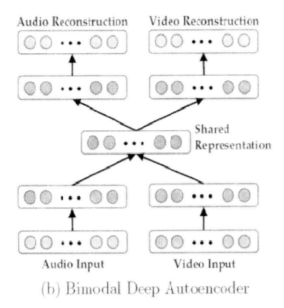

Figure 11.3: The architecture of a deep denoising autoencoder for multi-modal audio/speech and visual features. [after [269], @ICML].

learn a shared audio and video representation, and evaluate it on a fixed task, where the classifier is trained with audio-only data but tested with video-only data and vice versa. The work concludes that deep learning architectures are generally effective in learning multi-modal features from unlabeled data and in improving single modality features through cross modality information transfer. One exception is the cross-modality setting using the CUAVE dataset. The results presented in [269, 268] show that learning video features with both video and audio outperforms that with only video data. However, the same paper also shows that a model of [278] in which a sophisticated signal processing technique for extracting visual features, together with the uncertainty-compensation method developed originally from robust speech recognition [104], gives the best classification accuracy in the cross-modal learning task, beating the features derived from the generative deep architecture designed for this task.

While the deep generative architecture for multimodal learning described in [268, 269] is based on non-probabilistic autoencoder neural nets, a probabilistic version based on deep Boltzmann machine (DBM) has appeared more recently for the same multimodal application. In [348], a DBM is used to extract a unified representation integrating separate modalities, useful for both classification and information retrieval tasks. Rather than using the "bottleneck" layers in the deep autoencoder to represent multimodal inputs, here a probability density is defined on the joint space of multimodal inputs, and states of suitably defined latent variables are used for the representation. The advantage of this probabilistic formulation, possibly lacking in the traditional deep autoencoder, is that the missing modality's information can be filled in naturally by sampling from its conditional distribution. More recent work on autoencoders [22, 30] shows the capability of generalized denoising autoencoders in carrying out sampling, thus they may overcome the earlier problem of filling-in the missing modality's information. For the bi-modal data consisting of image and text, the multimodal DBM was shown to slightly outperform the traditional version of the deep multimodal autoencoder as well as multimodal DBN in classification and information retrieval tasks. No results on the comparisons with the generalized version of deep autoencoders has been reported but may appear soon.

The several architectures discussed so far in this chapter for multimodal processing and learning can be regarded as special cases of more general multi-task learning and transfer learning [22, 47]. Transfer learning, encompassing both adaptive and multi-task learning, refers to the ability of a learning architecture and technique to exploit common hidden explanatory factors among different learning tasks. Such exploitation permits sharing of aspects of diverse types of input data sets, thus allowing the possibility of transferring knowledge across seemingly different learning tasks. As argued in [22], the learning architecture shown in Figure 11.4 and the associated learning algorithms have an advantage for such tasks because they learn representations that capture underlying factors, a subset of which may be relevant for each particular task. We will discuss a number of such multi-task learning applications in the remainder of this chapter that are confined with a single modality of speech, natural language processing, *or* image domain.

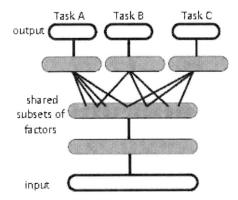

Figure 11.4: A DNN architecture for multitask learning that is aimed to discover hidden explanatory factors shared among three tasks A, B, and C. [after [22], @IEEE].

11.3 Multi-task learning within the speech, NLP or image domain

Within the speech domain, one most interesting application of multitask learning is multi-lingual or cross-lingual speech recognition, where speech recognition for different languages is considered as different tasks. Various approaches have been taken to attack this rather challenging acoustic modeling problem for speech recognition, where the difficulty lies in the lack of transcribed speech data due to economic considerations in developing speech recognition systems for all languages in the world. Cross-language data sharing and data weighing are common and useful approaches for the GMM–HMM system [225]. Another successful approach for the GMM–HMM is to map pronunciation units across languages either via knowledge-based or data-driven methods [420]. But they are much inferior to the DNN–HMM approach which we now summarize.

In recent papers of [94, 170] and [150], two research groups independently developed closely related DNN architectures with multi-task learning capabilities for multilingual speech recognition. See Figure 11.5 for an illustration of this type of architecture. The idea behind these architectures is that the hidden layers in the DNN, when learned

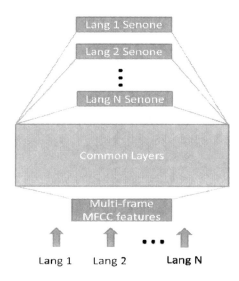

Figure 11.5: A DNN architecture for multilingual speech recognition. [after [170], @IEEE].

appropriately, serve as increasingly complex feature transformations sharing common hidden factors across the acoustic data in different languages. The final softmax layer representing a log-linear classifier makes use of the most abstract feature vectors represented in the topmost hidden layer. While the log-linear classifier is necessarily separate for different languages, the feature transformations can be shared across languages. Excellent multilingual speech recognition results are reported, far exceeding the earlier results using the GMM–HMM based approaches [225, 420]. The implication of this set of work is significant and far reaching. It points to the possibility of quickly building a high-performance DNN-based system for a new language from an existing multilingual DNN. This huge benefit would require only a small amount of training data from the target language, although having more data would further improve the performance. This multitask learning approach can reduce the need for the unsupervised pre-training stage, and can train the DNN with much fewer epochs. Extension of this set of work would be to efficiently build a language-universal speech recognition system. Such a system cannot only recognize many languages and improve the accuracy for each individual language, but

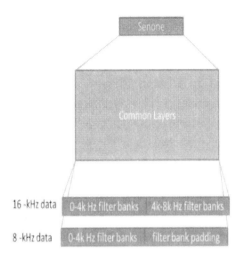

Figure 11.6: A DNN architecture for speech recognition trained with mixed-bandwidth acoustic data with 16-kHz and 8-kHz sampling rates; [after [221], @IEEE].

also expand the languages supported by simply stacking softmax layers on the DNN for new languages.

A closely related DNN architecture, as shown in Figure 11.6, with multitask learning capabilities was also recently applied to another acoustic modeling problem — learning joint representations for two separate sets of acoustic data [94, 221]. The set that consists of the speech data with 16 kHz sampling rate is of wideband and high quality, which is often collected from increasingly popular smart phones under the voice search scenario. Another, narrowband data set has a lower sampling rate of 8kHz, often collected using the telephony speech recognition systems.

As a final example of multi-task learning within the speech domain, let us consider phone recognition and word recognition as separate "tasks." That is, phone recognition results are used not for producing text outputs but for language-type identification or for spoken document retrieval. Then, the use of pronunciation dictionary in almost all speech systems can be considered as multi-task learning that share the tasks of phone recognition and word recognition. More advanced frameworks in speech recognition have pushed this direction further by advocating the use of even finer units of speech than phones to bridge the raw acoustic information of speech to semantic content of speech via a hierarchy of linguistic structure. These atomic speech units include "speech attributes" in the detection-based and knowledge-rich modeling framework for speech recognition, whose accuracy has been significantly boosted recently by the use of deep learning methods [332, 330, 427].

Within the natural language processing domain, the best known example of multi-task learning is the comprehensive studies reported in [62, 63], where a range of separate "tasks" of part-of-speech tagging, chunking, named entity tagging, semantic role identification, and similar-word identification in natural language processing are attacked using a common representation of words and a unified deep learning approach. A summary of these studies can be found in Section 8.2.

Finally, within the domain of image/vision as a single modality, deep learning has also been found effective in multi-task learning. Srivastava and Salakhutdinov [349] present a multi-task learning approach based on hierarchical Bayesian priors in a DNN system applied to various image classification data sets. The priors are combined with a DNN, which improves discriminative learning by encouraging information sharing among tasks and by discovering similar classes among which knowledge is transferred. More specifically, methods are developed to jointly learn to classify images and a hierarchy of classes, such that "poor classes," for which there are relatively few training examples, can benefit from similar "rich classes," for which more training examples are available. This work can be considered as an excellent instance of learning output representations, in addition to learning input representation of the DNN as the focus of nearly all deep learning work reported in the literature.

As another example of multi-task learning within the single-modality domain of image, Ciresan et al. [58] applied the architecture of deep CNNs to character recognition tasks for Latin and for Chinese. The deep CNNs trained on Chinese characters are shown to be easily capable of recognizing uppercase Latin letters. Further, learning Chinese characters is accelerated by first pre-training a CNN on a small subset of all classes and then continuing to train on all classes.

12

Conclusion

This monograph first presented a brief history of deep learning (focusing on speech recognition) and developed a categorization scheme to analyze the existing deep networks in the literature into unsupervised (many of which are generative), supervised, and hybrid classes. The deep autoencoder, the DSN (as well as many of its variants), and the DBN–DNN or pre-trained DNN architectures, one in each of the three classes, are discussed and analyzed in detail, as they appear to be popular and promising approaches based on the authors' personal research experiences. Applications of deep learning in five broad areas of information processing are also reviewed, including speech and audio (Section 7), natural language modeling and processing (Section 8), information retrieval (Section 9), object recognition and computer vision (Section 10), and multi-modal and multi-task learning (Section 11). There are other interesting yet non-mainstream applications of deep learning, which are not covered in this monograph. For interested readers, please consult recent papers on the applications of deep learning to optimal control in [219], to reinforcement learning in [256], to malware classification in [66], to compressed sensing in [277], to recognition confidence prediction in [173], to acoustic-articulatory inversion mapping in [369], to emotion recognition from video in [189], to emotion recognition from speech in [207, 222], to spoken language understanding in [242, 366, 403], to speaker recognition in [351, 372], to language-type recognition in [112], to dialogue state tracking for spoken dialogue systems in [94, 152], to automatic voice activity detection in [442], to speech enhancement in [396], to voice conversion in [266], and to single-channel source separation in [132, 387].

The literature on deep learning is vast, mostly coming from the machine learning community. The signal processing community embraced deep learning only within the past four years or so (starting around end of 2009) and the momentum is growing fast ever since. This monograph is written mainly from the signal and information processing perspective. Beyond surveying the existing deep learning work, a classificatory scheme based on the architectures and on the nature of the learning algorithms is developed, and an analysis and discussions with concrete examples are presented. It is our hope that the survey conducted in this monograph will provide insight for readers to better understand the capability of the various deep learning systems discussed in the monograph, the connection among different but similar deep learning methods, and how to design proper deep learning algorithms under different circumstances.

Throughout this review, the important message is conveyed that building and learning deep hierarchies of features are highly desirable. We have discussed the difficulty of learning parameters in all layers of deep networks in one shot due to optimization difficulties that need to be better understood. The unsupervised pre-training method in the hybrid architecture of the DBN–DNN, which we reviewed in detail in Section 5, appears to have offered a useful, albeit empirical, solution to poor local optima in optimization and to regularization for the deep model containing massive parameters even though a solid theoretical foundation is still lacking. The effectiveness of the pre-training method, which was one factor that stimulated the interest in deep learning by the signal processing community in 2009 via collaborations between academic and industrial researchers, is most prominent when the supervised training data are limited.

Deep learning is an emerging technology. Despite the empirical promising results reported so far, much more work needs to be carried out. Importantly, it has not been the experience of deep learning researchers that a single deep learning technique can be successful for all classification tasks. For example, while the popular learning strategy of generative pre-training followed by discriminative fine-tuning seems to work well empirically for many tasks, it failed to work for some other tasks that have been explored (e.g., language identification or speaker recognition; unpublished). For these tasks, the features extracted at the generative pre-training phase seem to describe the underlying speech variations well but do not contain sufficient information to distinguish between different languages. A learning strategy that can extract discriminative yet also invariant features is expected to provide better solutions. This idea has also been called "disentangling" and is developed further in [24]. Further, extracting discriminative features may greatly reduce the model size needed in many of the current deep learning systems. Domain knowledge such as what kind of invariance is useful for a specific task in hand (e.g., vision, speech, or natural language) and what kind of regularization in terms of parameter constraints is key to the success of applying deep learning methods. Moreover, new types of DNN architectures and learning beyond the several popular ones discussed in this monograph are currently under active development by the deep learning research community (e.g., [24, 89]), holding the promise to improve the performance of deep learning models in more challenging applications in signal processing and in artificial intelligence.

Recent published work showed that there is vast room to improve the current optimization techniques for learning deep architectures [69, 208, 238, 239, 311, 356, 393]. To what extent pre-training is essential to learning the full set of parameters in deep architectures is currently under investigation, especially when very large amounts of labeled training data are available, reducing or even obliterating the need for model regularization. Some preliminary results have been discussed in this monograph and in [55, 161, 323, 429].

In recent years, machine learning is becoming increasingly dependent on large-scale data sets. For instance, many of the recent successes of deep learning as discussed in this monograph have relied on the access to massive data sets and massive computing power. It would become increasingly difficult to explore the new algorithmic space without the access to large, real-world data sets and without the related engineering expertise. How well deep learning algorithms behave would depend heavily on the amount of data and computing power available. As we showed with speech recognition examples, a deep learning algorithm that appears to be performing not so remarkably on small data sets can begin to perform considerably better when these limitations are removed, one of main reasons for the recent resurgence in neural network research. As an example, the DBN pre-training that ignited a new era of (deep) machine learning research appears unnecessary if enough data and computing power are used.

As a consequence, effective and scalable parallel algorithms are critical for training deep models with large data sets, as in many common information processing applications such as speech recognition and machine translation. The popular mini-batch stochastic gradient technique is known to be difficult to parallelize over computers. The common practice nowadays is to use GPGPUs to speed up the learning process, although recent advance in developing asynchronous stochastic gradient descent learning has shown promises by using large-scale CPU clusters [69, 209] and GPU clusters [59]. In this interesting computing architecture, many different replicas of the DNN compute gradients on different subsets of the training data in parallel. These gradients are communicated to a central parameter server that updates the shared weights. Even though each replica typically computes gradients using parameter values not immediately updated, stochastic gradient descent is robust to the slight errors this has introduced. To make deep learning techniques scalable to very large training data, theoretically sound parallel learning and optimization algorithms together with novel architectures need to be further developed [31, 39, 49, 69, 181, 322, 356]. Optimization methods specific to speech recognition problems may need to be taken into account in order to push speech recognition advances to the next level [46, 149, 393].

One major barrier to the application of DNNs and related deep models is that it currently requires considerable skill and experience to choose sensible values for hyper-parameters such as the learning rate schedule, the strength of the regularizer, the number of layers and the number of units per layer, etc. Sensible values for one hyper-parameter may depend on the values chosen for other hyper-parameters and hyper-parameter tuning in DNNs is especially expensive. Some interesting methods for solving the problem have been developed recently, including random sampling [32] and Bayesian optimization procedure [337]. Further research is needed in this important area.

This monograph, mainly in Sections 8 and 11 on natural language and multi-modal applications, has touched on some recent work on using deep learning methods to do reasoning, moving beyond the topic of more straightforward pattern recognition using supervised, unsupervised or hybrid learning methods to which much of this monograph has been devoted to. In principle, since deep networks are naturally equipped with distributed representations (rf. Table 3.1) using their layer-wise collections of units for coding relations and coding entities, concepts, events, topics, etc., they can potentially perform powerful reasoning over structures, as argued in various historical publications as well as recent essays [38, 156, 286, 288, 292, 336, 335]. While initial explorations on this capability of deep networks have recently appeared in the literature, as reviewed in Sections 8 and 11, much research is needed. If successful, this new type of deep learning "machine" will open up many novel and exciting applications in applied artificial intelligence as a "thinking brain." We expect growing work of deep learning in this area, full of new challenges, in the future.

Further, solid theoretical foundations of deep learning need to be established in a myriad of aspects. As an example, the success of deep learning in unsupervised learning has not been demonstrated as much as for supervised learning; yet the essence and major motivation of deep learning lie right in unsupervised learning for automatically discovering data representation. The issues involve appropriate objectives for learning effective feature representations and the right deep learning architectures/algorithms for distributed representations to effectively disentangle the hidden explanatory factors of variation in the data. Unfortunately, a majority of the successful deep learning techniques have so far dealt with unstructured or "flat" classification problems. For example, although speech recognition is a sequential classification problem by nature, in the most successful and large-scale systems, a separate HMM is used to handle the sequence structure and the DNN is only used to produce the frame-level, unstructured posterior distributions. Recent proposals have called for and investigated moving beyond the "flat" representations and incorporating structures in both the deep learning architectures and input and output representations [79, 136, 338, 349].

Finally, deep learning researchers have been advised by neuroscientists to seriously consider a broader set of issues and learning architectures so as to gain insight into biologically plausible representations in the brain that may be useful for practical applications [272]. How can computational neuroscience models about hierarchical brain structure help improve engineering deep learning architectures? How may the biologically feasible learning styles in the brain [158, 395] help design more effective and more robust deep learning algorithms? All these issues and those discussed earlier in this section will need intensive research in order to further push the frontier of deep learning.

References

[1] O. Abdel-Hamid, L. Deng, and D. Yu. Exploring convolutional neural network structures and optimization for speech recognition. *Proceedings of Interspeech*, 2013.

[2] O. Abdel-Hamid, L. Deng, D. Yu, and H. Jiang. Deep segmental neural networks for speech recognition. In *Proceedings of Interspeech*. 2013.

[3] O. Abdel-Hamid, A. Mohamed, H. Jiang, and G. Penn. Applying convolutional neural networks concepts to hybrid NN-HMM model for speech recognition. In *Proceedings of International Conference on Acoustics Speech and Signal Processing (ICASSP)*. 2012.

[4] A. Acero, L. Deng, T. Kristjansson, and J. Zhang. HMM adaptation using vector taylor series for noisy speech recognition. In *Proceedings of Interspeech*. 2000.

[5] G. Alain and Y. Bengio. What regularized autoencoders learn from the data generating distribution. In *Proceedings of International Conference on Learning Representations (ICLR)*. 2013.

[6] G. Anthes. Deep learning comes of age. *Communications of the Association for Computing Machinery (ACM)*, 56(6):13–15, June 2013.

[7] I. Arel, C. Rose, and T. Karnowski. Deep machine learning — a new frontier in artificial intelligence. *IEEE Computational Intelligence Magazine*, 5:13–18, November 2010.

[8] E. Arisoy, T. Sainath, B. Kingsbury, and B. Ramabhadran. Deep neural network language models. In *Proceedings of the Joint Human Language Technology Conference and the North American Chapter of the Association of Computational Linguistics (HLT-NAACL) Workshop*. 2012.

[9] O. Aslan, H. Cheng, D. Schuurmans, and X. Zhang. Convex two-layer modeling. In *Proceedings of Neural Information Processing Systems (NIPS)*. 2013.

[10] J. Ba and B. Frey. Adaptive dropout for training deep neural networks. In *Proceedings of Neural Information Processing Systems (NIPS)*. 2013.

[11] J. Baker, L. Deng, J. Glass, S. Khudanpur, C.-H. Lee, N. Morgan, and D. O'Shaughnessy. Research developments and directions in speech recognition and understanding. *IEEE Signal Processing Magazine*, 26(3):75–80, May 2009.

[12] J. Baker, L. Deng, J. Glass, S. Khudanpur, C.-H. Lee, N. Morgan, and D. O'Shaughnessy. Updated MINS report on speech recognition and understanding. *IEEE Signal Processing Magazine*, 26(4), July 2009.

[13] P. Baldi and P. Sadowski. Understanding dropout. In *Proceedings of Neural Information Processing Systems (NIPS)*. 2013.

[14] E. Battenberg, E. Schmidt, and J. Bello. *Deep learning for music, special session at International Conference on Acoustics Speech and Signal Processing (ICASSP)* (http://www.icassp2014.org/special_sections.html#ss8), 2014.

[15] E. Batternberg and D. Wessel. Analyzing drum patterns using conditional deep belief networks. In *Proceedings of International Symposium on Music Information Retrieval (ISMIR)*. 2012.

[16] P. Bell, P. Swietojanski, and S. Renals. Multi-level adaptive networks in tandem and hybrid ASR systems. In *Proceedings of International Conference on Acoustics Speech and Signal Processing (ICASSP)*. 2013.

[17] Y. Bengio. Artificial neural networks and their application to sequence recognition. Ph.D. Thesis, McGill University, Montreal, Canada, 1991.

[18] Y. Bengio. New distributed probabilistic language models. Technical Report, University of Montreal, 2002.

[19] Y. Bengio. Neural net language models. *Scholarpedia*, 3, 2008.

[20] Y. Bengio. Learning deep architectures for AI. in *Foundations and Trends in Machine Learning*, 2(1):1–127, 2009.

[21] Y. Bengio. Deep learning of representations for unsupervised and transfer learning. *Journal of Machine Learning Research Workshop and Conference Proceedings*, 27:17–37, 2012.

[22] Y. Bengio. Deep learning of representations: Looking forward. In *Statistical Language and Speech Processing*, pages 1–37. Springer, 2013.

[23] Y. Bengio, N. Boulanger, and R. Pascanu. Advances in optimizing recurrent networks. In *Proceedings of International Conference on Acoustics Speech and Signal Processing (ICASSP)*. 2013.

[24] Y. Bengio, A. Courville, and P. Vincent. Representation learning: A review and new perspectives. *IEEE Transactions on Pattern Analysis and Machine Intelligence (PAMI)*, 38:1798–1828, 2013.

[25] Y. Bengio, R. De Mori, G. Flammia, and R. Kompe. Global optimization of a neural network-hidden markov model hybrid. *IEEE Transactions on Neural Networks*, 3:252–259, 1992.

[26] Y. Bengio, R. Ducharme, P. Vincent, and C. Jauvin. A neural probabilistic language model. In *Proceedings of Neural Information Processing Systems (NIPS)*. 2000.

[27] Y. Bengio, R. Ducharme, P. Vincent, and C. Jauvin. A neural probabilistic language model. *Journal of Machine Learning Research*, 3:1137–1155, 2003.

[28] Y. Bengio, P. Lamblin, D. Popovici, and H. Larochelle. Greedy layerwise training of deep networks. In *Proceedings of Neural Information Processing Systems (NIPS)*. 2006.

[29] Y. Bengio, P. Simard, and P. Frasconi. Learning long-term dependencies with gradient descent is difficult. *IEEE Transactions on Neural Networks*, 5:157–166, 1994.

[30] Y. Bengio, E. Thibodeau-Laufer, and J. Yosinski. Deep generative stochastic networks trainable by backprop. arXiv 1306:1091, 2013. also accepted to appear in *Proceedings of International Conference on Machine Learning (ICML), 2014*.

[31] Y. Bengio, L. Yao, G. Alain, and P. Vincent. Generalized denoising autoencoders as generative models. In *Proceedings of Neural Information Processing Systems (NIPS)*. 2013.

[32] J. Bergstra and Y. Bengio. Random search for hyper-parameter optimization. *Journal on Machine Learning Research*, 3:281–305, 2012.

[33] A. Biem, S. Katagiri, E. McDermott, and B. Juang. An application of discriminative feature extraction to filter-bank-based speech recognition. *IEEE Transactions on Speech and Audio Processing*, 9:96–110, 2001.

[34] J. Bilmes. Dynamic graphical models. *IEEE Signal Processing Magazine*, 33:29–42, 2010.

[35] J. Bilmes and C. Bartels. Graphical model architectures for speech recognition. *IEEE Signal Processing Magazine*, 22:89–100, 2005.

[36] A. Bordes, X. Glorot, J. Weston, and Y. Bengio. A semantic matching energy function for learning with multi-relational data — application to word-sense disambiguation. *Machine Learning*, May 2013.

[37] A. Bordes, J. Weston, R. Collobert, and Y. Bengio. Learning structured embeddings of knowledge bases. In *Proceedings of Association for the Advancement of Artificial Intelligence (AAAI)*. 2011.

[38] L. Bottou. From machine learning to machine reasoning: An essay. *Journal of Machine Learning Research*, 14:3207–3260, 2013.

[39] L. Bottou and Y. LeCun. Large scale online learning. In *Proceedings of Neural Information Processing Systems (NIPS)*. 2004.

[40] N. Boulanger-Lewandowski, Y. Bengio, and P. Vincent. Modeling Temporal dependencies in high-dimensional sequences: Application to polyphonic music generation and transcription. In *Proceedings of International Conference on Machine Learning (ICML)*. 2012.

[41] N. Boulanger-Lewandowski, Y. Bengio, and P. Vincent. Audio chord recognition with recurrent neural networks. In *Proceedings of International Symposium on Music Information Retrieval (ISMIR)*. 2013.

[42] H. Bourlard and N. Morgan. *Connectionist Speech Recognition: A Hybrid Approach*. Kluwer, Norwell, MA, 1993.

[43] J. Bouvrie. Hierarchical learning: Theory with applications in speech and vision. Ph.D. thesis, MIT, 2009.

[44] L. Breiman. Stacked regression. *Machine Learning*, 24:49–64, 1996.

[45] J. Bridle, L. Deng, J. Picone, H. Richards, J. Ma, T. Kamm, M. Schuster, S. Pike, and R. Reagan. An investigation of segmental hidden dynamic models of speech coarticulation for automatic speech recognition. Final Report for 1998 Workshop on Language Engineering, CLSP, Johns Hopkins, 1998.

[46] P. Cardinal, P. Dumouchel, and G. Boulianne. Large vocabulary speech recognition on parallel architectures. *IEEE Transactions on Audio, Speech, and Language Processing*, 21(11):2290–2300, November 2013.

[47] R. Caruana. Multitask learning. *Machine Learning*, 28:41–75, 1997.

[48] J. Chen and L. Deng. A primal-dual method for training recurrent neural networks constrained by the echo-state property. In *Proceedings of International Conference on Learning Representations*. April 2014.

[49] X. Chen, A. Eversole, G. Li, D. Yu, and F. Seide. Pipelined back-propagation for context-dependent deep neural networks. In *Proceedings of Interspeech*. 2012.

[50] R. Chengalvarayan and L. Deng. Hmm-based speech recognition using state-dependent, discriminatively derived transforms on Mel-warped DFT features. *IEEE Transactions on Speech and Audio Processing*, pages 243–256, 1997.

[51] R. Chengalvarayan and L. Deng. Use of generalized dynamic feature parameters for speech recognition. *IEEE Transactions on Speech and Audio Processing*, pages 232–242, 1997a.

[52] R. Chengalvarayan and L. Deng. Speech trajectory discrimination using the minimum classification error learning. *IEEE Transactions on Speech and Audio Processing*, 6(6):505–515, 1998.

[53] Y. Cho and L. Saul. Kernel methods for deep learning. In *Proceedings of Neural Information Processing Systems (NIPS)*, pages 342–350. 2009.

[54] D. Ciresan, A. Giusti, L. Gambardella, and J. Schmidhuber. Deep neural networks segment neuronal membranes in electron microscopy images. In *Proceedings of Neural Information Processing Systems (NIPS)*. 2012.

[55] D. Ciresan, U. Meier, L. Gambardella, and J. Schmidhuber. Deep, big, simple neural nets for handwritten digit recognition. *Neural Computation*, December 2010.

[56] D. Ciresan, U. Meier, J. Masci, and J. Schmidhuber. A committee of neural networks for traffic sign classification. In *Proceedings of International Joint Conference on Neural Networks (IJCNN)*. 2011.

[57] D. Ciresan, U. Meier, and J. Schmidhuber. Multi-column deep neural networks for image classification. In *Proceedings of Computer Vision and Pattern Recognition (CVPR)*. 2012.

[58] D. C. Ciresan, U. Meier, and J. Schmidhuber. Transfer learning for Latin and Chinese characters with deep neural networks. In *Proceedings of International Joint Conference on Neural Networks (IJCNN)*. 2012.

[59] A. Coates, B. Huval, T. Wang, D. Wu, A. Ng, and B. Catanzaro. Deep learning with COTS HPC. In *Proceedings of International Conference on Machine Learning (ICML)*. 2013.

[60] W. Cohen and R. V. de Carvalho. Stacked sequential learning. In *Proceedings of International Joint Conference on Artificial Intelligence (IJCAI)*, pages 671–676. 2005.

[61] R. Collobert. Deep learning for efficient discriminative parsing. In *Proceedings of Artificial Intelligence and Statistics (AISTATS)*. 2011.

[62] R. Collobert and J. Weston. A unified architecture for natural language processing: Deep neural networks with multitask learning. In *Proceedings of International Conference on Machine Learning (ICML)*. 2008.

[63] R. Collobert, J. Weston, L. Bottou, M. Karlen, K. Kavukcuoglu, and P. Kuksa. Natural language processing (almost) from scratch. *Journal on Machine Learning Research*, 12:2493–2537, 2011.

[64] G. Dahl, M. Ranzato, A. Mohamed, and G. Hinton. Phone recognition with the mean-covariance restricted boltzmann machine. In *Proceedings of Neural Information Processing Systems (NIPS)*, volume 23, pages 469–477. 2010.

[65] G. Dahl, T. Sainath, and G. Hinton. Improving deep neural networks for LVCSR using rectified linear units and dropout. In *Proceedings of International Conference on Acoustics Speech and Signal Processing (ICASSP)*. 2013.

[66] G. Dahl, J. Stokes, L. Deng, and D. Yu. Large-scale malware classification using random projections and neural networks. In *Proceedings of International Conference on Acoustics Speech and Signal Processing (ICASSP)*. 2013.

[67] G. Dahl, D. Yu, L. Deng, and A. Acero. Context-dependent DBN-HMMs in large vocabulary continuous speech recognition. In *Proceedings of International Conference on Acoustics Speech and Signal Processing (ICASSP)*. 2011.

[68] G. Dahl, D. Yu, L. Deng, and A. Acero. Context-dependent, pre-trained deep neural networks for large vocabulary speech recognition. *IEEE Transactions on Audio, Speech, & Language Processing*, 20(1):30–42, January 2012.

[69] J. Dean, G. Corrado, R. Monga, K. Chen, M. Devin, Q. Le, M. Mao, M. Ranzato, A. Senior, P. Tucker, K. Yang, and A. Ng. Large scale distributed deep networks. In *Proceedings of Neural Information Processing Systems (NIPS)*. 2012.

[70] K. Demuynck and F. Triefenbach. Porting concepts from DNNs back to GMMs. In *Proceedings of the Automatic Speech Recognition and Understanding Workshop (ASRU)*. 2013.

[71] L. Deng. A generalized hidden Markov model with state-conditioned trend functions of time for the speech signal. *Signal Processing*, 27(1):65–78, 1992.

[72] L. Deng. A stochastic model of speech incorporating hierarchical nonstationarity. *IEEE Transactions on Speech and Audio Processing*, 1(4):471–475, 1993.

[73] L. Deng. A dynamic, feature-based approach to the interface between phonology and phonetics for speech modeling and recognition. *Speech Communication*, 24(4):299–323, 1998.

[74] L. Deng. Computational models for speech production. In *Computational Models of Speech Pattern Processing*, pages 199–213. Springer Verlag, 1999.

[75] L. Deng. Switching dynamic system models for speech articulation and acoustics. In *Mathematical Foundations of Speech and Language Processing*, pages 115–134. Springer-Verlag, New York, 2003.

[76] L. Deng. *Dynamic Speech Models — Theory, Algorithm, and Application*. Morgan & Claypool, December 2006.

[77] L. Deng. An overview of deep-structured learning for information processing. In *Proceedings of Asian-Pacific Signal & Information Processing Annual Summit and Conference (APSIPA-ASC)*. October 2011.

[78] L. Deng. The MNIST database of handwritten digit images for machine learning research. *IEEE Signal Processing Magazine*, 29(6), November 2012.

[79] L. Deng. Design and learning of output representations for speech recognition. In *Neural Information Processing Systems (NIPS) Workshop on Learning Output Representations*. December 2013.

[80] L. Deng. A tutorial survey of architectures, algorithms, and applications for deep learning. In *Asian-Pacific Signal & Information Processing Association Transactions on Signal and Information Processing*. 2013.

[81] L. Deng, O. Abdel-Hamid, and D. Yu. A deep convolutional neural network using heterogeneous pooling for trading acoustic invariance with phonetic confusion. In *Proceedings of International Conference on Acoustics Speech and Signal Processing (ICASSP)*. 2013.

[82] L. Deng, A. Acero, L. Jiang, J. Droppo, and X. Huang. High performance robust speech recognition using stereo training data. In *Proceedings of International Conference on Acoustics Speech and Signal Processing (ICASSP)*. 2001.

[83] L. Deng and M. Aksmanovic. Speaker-independent phonetic classification using hidden markov models with state-conditioned mixtures of trend functions. *IEEE Transactions on Speech and Audio Processing*, 5:319–324, 1997.

[84] L. Deng, M. Aksmanovic, D. Sun, and J. Wu. Speech recognition using hidden Markov models with polynomial regression functions as nonstationary states. *IEEE Transactions on Speech and Audio Processing*, 2(4):507–520, 1994.

[85] L. Deng and J. Chen. Sequence classification using the high-level features extracted from deep neural networks. In *Proceedings of International Conference on Acoustics Speech and Signal Processing (ICASSP)*. 2014.

[86] L. Deng and K. Erler. Structural design of a hidden Markov model based speech recognizer using multi-valued phonetic features: Comparison with segmental speech units. *Journal of the Acoustical Society of America*, 92(6):3058–3067, 1992.

[87] L. Deng, K. Hassanein, and M. Elmasry. Analysis of correlation structure for a neural predictive model with application to speech recognition. *Neural Networks*, 7(2):331–339, 1994.

[88] L. Deng, X. He, and J. Gao. Deep stacking networks for information retrieval. In *Proceedings of International Conference on Acoustics Speech and Signal Processing (ICASSP)*. 2013c.

[89] L. Deng, G. Hinton, and B. Kingsbury. New types of deep neural network learning for speech recognition and related applications: An overview. In *Proceedings of International Conference on Acoustics Speech and Signal Processing (ICASSP)*. 2013b.

[90] L. Deng and X. D. Huang. Challenges in adopting speech recognition. *Communications of the Association for Computing Machinery (ACM)*, 47(1):11–13, January 2004.

[91] L. Deng, B. Hutchinson, and D. Yu. Parallel training of deep stacking networks. In *Proceedings of Interspeech*. 2012b.

[92] L. Deng, M. Lennig, V. Gupta, F. Seitz, P. Mermelstein, and P. Kenny. Phonemic hidden Markov models with continuous mixture output densities for large vocabulary word recognition. *IEEE Transactions on Signal Processing*, 39(7):1677–1681, 1991.

[93] L. Deng, M. Lennig, F. Seitz, and P. Mermelstein. Large vocabulary word recognition using context-dependent allophonic hidden Markov models. *Computer Speech and Language*, 4(4):345–357, 1990.

[94] L. Deng, J. Li, K. Huang, Yao, D. Yu, F. Seide, M. Seltzer, G. Zweig, X. He, J. Williams, Y. Gong, and A. Acero. Recent advances in deep learning for speech research at Microsoft. In *Proceedings of International Conference on Acoustics Speech and Signal Processing (ICASSP)*. 2013a.

[95] L. Deng and X. Li. Machine learning paradigms in speech recognition: An overview. *IEEE Transactions on Audio, Speech, & Language*, 21:1060–1089, May 2013.

[96] L. Deng and J. Ma. Spontaneous speech recognition using a statistical coarticulatory model for the vocal tract resonance dynamics. *Journal of the Acoustical Society America*, 108:3036–3048, 2000.

[97] L. Deng and D. O'Shaughnessy. *Speech Processing — A Dynamic and Optimization-Oriented Approach*. Marcel Dekker, 2003.

[98] L. Deng, G. Ramsay, and D. Sun. Production models as a structural basis for automatic speech recognition. *Speech Communication*, 33(2–3):93–111, August 1997.

[99] L. Deng and H. Sameti. Transitional speech units and their representation by regressive Markov states: Applications to speech recognition. *IEEE Transactions on speech and audio processing*, 4(4):301–306, July 1996.

[100] L. Deng, M. Seltzer, D. Yu, A. Acero, A. Mohamed, and G. Hinton. Binary coding of speech spectrograms using a deep autoencoder. In *Proceedings of Interspeech*. 2010.

[101] L. Deng and D. Sun. A statistical approach to automatic speech recognition using the atomic speech units constructed from overlapping articulatory features. *Journal of the Acoustical Society of America*, 85(5):2702–2719, 1994.

[102] L. Deng, G. Tur, X. He, and D. Hakkani-Tur. Use of kernel deep convex networks and end-to-end learning for spoken language understanding. In *Proceedings of IEEE Workshop on Spoken Language Technologies*. December 2012.

[103] L. Deng, K. Wang, A. Acero, H. W. Hon, J. Droppo, C. Boulis, Y. Wang, D. Jacoby, M. Mahajan, C. Chelba, and X. Huang. Distributed speech processing in mipad's multimodal user interface. *IEEE Transactions on Speech and Audio Processing*, 10(8):605–619, 2002.

[104] L. Deng, J. Wu, J. Droppo, and A. Acero. Dynamic compensation of HMM variances using the feature enhancement uncertainty computed from a parametric model of speech distortion. *IEEE Transactions on Speech and Audio Processing*, 13(3):412–421, 2005.

[105] L. Deng and D. Yu. Use of differential cepstra as acoustic features in hidden trajectory modeling for phonetic recognition. In *Proceedings of International Conference on Acoustics Speech and Signal Processing (ICASSP)*. 2007.

[106] L. Deng and D. Yu. Deep convex network: A scalable architecture for speech pattern classification. In *Proceedings of Interspeech*. 2011.

[107] L. Deng, D. Yu, and A. Acero. A bidirectional target filtering model of speech coarticulation: Two-stage implementation for phonetic recognition. *IEEE Transactions on Audio and Speech Processing*, 14(1):256–265, January 2006.

[108] L. Deng, D. Yu, and A. Acero. Structured speech modeling. *IEEE Transactions on Audio, Speech and Language Processing*, 14(5):1492–1504, September 2006.

[109] L. Deng, D. Yu, and G. Hinton. Deep learning for speech recognition and related applications. *Neural Information Processing Systems (NIPS) Workshop*, 2009.

[110] L. Deng, D. Yu, and J. Platt. Scalable stacking and learning for building deep architectures. In *Proceedings of International Conference on Acoustics Speech and Signal Processing (ICASSP)*. 2012a.

[111] T. Deselaers, S. Hasan, O. Bender, and H. Ney. A deep learning approach to machine transliteration. In *Proceedings of 4th Workshop on Statistical Machine Translation*, pages 233–241. Athens, Greece, March 2009.

[112] A. Diez. Automatic language recognition using deep neural networks. Thesis, Universidad Autonoma de Madrid, SPAIN, September 2013.

[113] P. Dognin and V. Goel. Combining stochastic average gradient and hessian-free optimization for sequence training of deep neural networks. In *Proceedings of the Automatic Speech Recognition and Understanding Workshop (ASRU)*. 2013.

[114] D. Erhan, Y. Bengio, A. Courvelle, P. Manzagol, P. Vencent, and S. Bengio. Why does unsupervised pre-training help deep learning? *Journal on Machine Learning Research*, pages 201–208, 2010.

[115] R. Fernandez, A. Rendel, B. Ramabhadran, and R. Hoory. F0 contour prediction with a deep belief network-gaussian process hybrid model. In *Proceedings of International Conference on Acoustics Speech and Signal Processing (ICASSP)*, pages 6885–6889. 2013.

[116] S. Fine, Y. Singer, and N. Tishby. The hierarchical hidden Markov model: Analysis and applications. *Machine Learning*, 32:41–62, 1998.

[117] A. Frome, G. Corrado, J. Shlens, S. Bengio, J. Dean, M. Ranzato, and T. Mikolov. Devise: A deep visual-semantic embedding model. In *Proceedings of Neural Information Processing Systems (NIPS)*. 2013.

[118] Q. Fu, X. He, and L. Deng. Phone-discriminating minimum classification error (p-mce) training for phonetic recognition. In *Proceedings of Interspeech*. 2007.

[119] M. Gales. Model-based approaches to handling uncertainty. In *Robust Speech Recognition of Uncertain or Missing Data: Theory and Application*, pages 101–125. Springer, 2011.

[120] J. Gao, X. He, and J.-Y. Nie. Clickthrough-based translation models for web search: From word models to phrase models. In *Proceedings of Conference on Information and Knowledge Management (CIKM)*. 2010.

[121] J. Gao, X. He, W. Yih, and L. Deng. Learning semantic representations for the phrase translation model. In *Proceedings of Neural Information Processing Systems (NIPS) Workshop on Deep Learning*. December 2013.

[122] J. Gao, X. He, W. Yih, and L. Deng. Learning semantic representations for the phrase translation model. MSR-TR-2013-88, September 2013.

[123] J. Gao, X. He, W. Yih, and L. Deng. Learning continuous phrase representations for translation modeling. In *Proceedings of Association for Computational Linguistics (ACL)*. 2014.

[124] J. Gao, K. Toutanova, and W.-T. Yih. Clickthrough-based latent semantic models for web search. In *Proceedings of Special Interest Group on Information Retrieval (SIGIR)*. 2011.

[125] R. Gens and P. Domingo. Discriminative learning of sum-product networks. *Neural Information Processing Systems (NIPS)*, 2012.

[126] D. George. How the brain might work: A hierarchical and temporal model for learning and recognition. Ph.D. thesis, Stanford University, 2008.

[127] M. Gibson and T. Hain. Error approximation and minimum phone error acoustic model estimation. *IEEE Transactions on Audio, Speech, and Language Processing*, 18(6):1269–1279, August 2010.

[128] R. Girshick, J. Donahue, T. Darrell, and J. Malik. Rich feature hierarchies for accurate object detection and semantic segmentation. arXiv:1311.2524v1, 2013.

[129] X. Glorot and Y. Bengio. Understanding the difficulty of training deep feed-forward neural networks. In *Proceedings of Artificial Intelligence and Statistics (AISTATS)*. 2010.

[130] X. Glorot, A. Bordes, and Y. Bengio. Deep sparse rectifier neural networks. In *Proceedings of Artificial Intelligence and Statistics (AIS-TATS)*. April 2011.

[131] I. Goodfellow, M. Mirza, A. Courville, and Y. Bengio. Multi-prediction deep boltzmann machines. In *Proceedings of Neural Information Processing Systems (NIPS)*. 2013.

[132] E. Grais, M. Sen, and H. Erdogan. Deep neural networks for single channel source separation. arXiv:1311.2746v1, 2013.

[133] A. Graves. Sequence transduction with recurrent neural networks. *Representation Learning Workshop, International Conference on Machine Learning (ICML)*, 2012.

[134] A. Graves, S. Fernandez, F. Gomez, and J. Schmidhuber. Connectionist temporal classification: Labeling unsegmented sequence data with recurrent neural networks. In *Proceedings of International Conference on Machine Learning (ICML)*. 2006.

[135] A. Graves, N. Jaitly, and A. Mohamed. Hybrid speech recognition with deep bidirectional LSTM. In *Proceedings of the Automatic Speech Recognition and Understanding Workshop (ASRU)*. 2013.

[136] A. Graves, A. Mohamed, and G. Hinton. Speech recognition with deep recurrent neural networks. In *Proceedings of International Conference on Acoustics Speech and Signal Processing (ICASSP)*. 2013.

[137] F. Grezl and P. Fousek. Optimizing bottle-neck features for LVCSR. In *Proceedings of International Conference on Acoustics Speech and Signal Processing (ICASSP)*. 2008.

[138] C. Gulcehre, K. Cho, R. Pascanu, and Y. Bengio. Learned-norm pooling for deep feedforward and recurrent neural networks. http://arxiv.org/abs/1311.1780, 2014.

[139] M. Gutmann and A. Hyvarinen. Noise-contrastive estimation of unnormalized statistical models, with applications to natural image statistics. *Journal of Machine Learning Research*, 13:307–361, 2012.

[140] T. Hain, L. Burget, J. Dines, P. Garner, F. Grezl, A. Hannani, M. Huijbregts, M. Karafiat, M. Lincoln, and V. Wan. Transcribing meetings with the AMIDA systems. *IEEE Transactions on Audio, Speech, and Language Processing*, 20:486–498, 2012.

[141] P. Hamel and D. Eck. Learning features from music audio with deep belief networks. In *Proceedings of International Symposium on Music Information Retrieval (ISMIR)*. 2010.

[142] G. Hawkins, S. Ahmad, and D. Dubinsky. Hierarchical temporal memory including HTM cortical learning algorithms. Numenta Technical Report, December 10 2010.

[143] J. Hawkins and S. Blakeslee. *On Intelligence: How a New Understanding of the Brain will lead to the Creation of Truly Intelligent Machines.* Times Books, New York, 2004.

[144] X. He and L. Deng. Speech recognition, machine translation, and speech translation — a unifying discriminative framework. *IEEE Signal Processing Magazine*, 28, November 2011.

[145] X. He and L. Deng. Optimization in speech-centric information processing: Criteria and techniques. In *Proceedings of International Conference on Acoustics Speech and Signal Processing (ICASSP)*. 2012.

[146] X. He and L. Deng. Speech-centric information processing: An optimization-oriented approach. In *Proceedings of the IEEE*. 2013.

[147] X. He, L. Deng, and W. Chou. Discriminative learning in sequential pattern recognition — a unifying review for optimization-oriented speech recognition. *IEEE Signal Processing Magazine*, 25:14–36, 2008.

[148] G. Heigold, H. Ney, P. Lehnen, T. Gass, and R. Schluter. Equivalence of generative and log-liner models. *IEEE Transactions on Audio, Speech, and Language Processing*, 19(5):1138–1148, February 2011.

[149] G. Heigold, H. Ney, and R. Schluter. Investigations on an EM-style optimization algorithm for discriminative training of HMMs. *IEEE Transactions on Audio, Speech, and Language Processing*, 21(12):2616–2626, December 2013.

[150] G. Heigold, V. Vanhoucke, A. Senior, P. Nguyen, M. Ranzato, M. Devin, and J. Dean. Multilingual acoustic models using distributed deep neural networks. In *Proceedings of International Conference on Acoustics Speech and Signal Processing (ICASSP)*. 2013.

[151] I. Heintz, E. Fosler-Lussier, and C. Brew. Discriminative input stream combination for conditional random field phone recognition. *IEEE Transactions on Audio, Speech, and Language Processing*, 17(8):1533–1546, November 2009.

[152] M. Henderson, B. Thomson, and S. Young. Deep neural network approach for the dialog state tracking challenge. In *Proceedings of Special Interest Group on Disclosure and Dialogue (SIGDIAL)*. 2013.

[153] M. Hermans and B. Schrauwen. Training and analysing deep recurrent neural networks. In *Proceedings of Neural Information Processing Systems (NIPS)*. 2013.

[154] H. Hermansky, D. Ellis, and S. Sharma. Tandem connectionist feature extraction for conventional HMM systems. In *Proceedings of International Conference on Acoustics Speech and Signal Processing (ICASSP)*. 2000.

[155] Y. Hifny and S. Renals. Speech recognition using augmented conditional random fields. *IEEE Transactions on Audio, Speech, and Language Processing*, 17(2):354–365, February 2009.

[156] G. Hinton. Mapping part-whole hierarchies into connectionist networks. *Artificial Intelligence*, 46:47–75, 1990.

[157] G. Hinton. Preface to the special issue on connectionist symbol processing. *Artificial Intelligence*, 46:1–4, 1990.

[158] G. Hinton. The ups and downs of Hebb synapses. *Canadian Psychology*, 44:10–13, 2003.

[159] G. Hinton. A practical guide to training restricted boltzmann machines. UTML Tech Report 2010-003, Univ. Toronto, August 2010.

[160] G. Hinton. A better way to learn features. *Communications of the Association for Computing Machinery (ACM)*, 54(10), October 2011.

[161] G. Hinton, L. Deng, D. Yu, G. Dahl, A. Mohamed, N. Jaitly, A. Senior, V. Vanhoucke, P. Nguyen, T. Sainath, and B. Kingsbury. Deep neural networks for acoustic modeling in speech recognition. *IEEE Signal Processing Magazine*, 29(6):82–97, November 2012.

[162] G. Hinton, A. Krizhevsky, and S. Wang. Transforming autoencoders. In *Proceedings of International Conference on Artificial Neural Networks*. 2011.

[163] G. Hinton, S. Osindero, and Y. Teh. A fast learning algorithm for deep belief nets. *Neural Computation*, 18:1527–1554, 2006.

[164] G. Hinton and R. Salakhutdinov. Reducing the dimensionality of data with neural networks. *Science*, 313(5786):504–507, July 2006.

[165] G. Hinton and R. Salakhutdinov. Discovering binary codes for documents by learning deep generative models. *Topics in Cognitive Science*, pages 1–18, 2010.

[166] G. Hinton, N. Srivastava, A. Krizhevsky, I. Sutskever, and R. Salakhutdinov. Improving neural networks by preventing co-adaptation of feature detectors. arXiv: 1207.0580v1, 2012.

[167] S. Hochreiter. Untersuchungen zu dynamischen neuronalen netzen. Diploma thesis, Institut fur Informatik, Technische Universitat Munchen, 1991.

[168] S. Hochreiter and J. Schmidhuber. Long short-term memory. *Neural Computation*, 9:1735–1780, 1997.

[169] E. Huang, R. Socher, C. Manning, and A. Ng. Improving word representations via global context and multiple word prototypes. In *Proceedings of Association for Computational Linguistics (ACL)*. 2012.

[170] J. Huang, J. Li, L. Deng, and D. Yu. Cross-language knowledge transfer using multilingual deep neural networks with shared hidden layers. In *Proceedings of International Conference on Acoustics Speech and Signal Processing (ICASSP)*. 2013.

[171] P. Huang, L. Deng, M. Hasegawa-Johnson, and X. He. Random features for kernel deep convex network. In *Proceedings of International Conference on Acoustics Speech and Signal Processing (ICASSP)*. 2013.

[172] P. Huang, X. He, J. Gao, L. Deng, A. Acero, and L. Heck. Learning deep structured semantic models for web search using clickthrough data. *Association for Computing Machinery (ACM) International Conference Information and Knowledge Management (CIKM)*, 2013.

[173] P. Huang, K. Kumar, C. Liu, Y. Gong, and L. Deng. Predicting speech recognition confidence using deep learning with word identity and score features. In *Proceedings of International Conference on Acoustics Speech and Signal Processing (ICASSP)*. 2013.

[174] S. Huang and S. Renals. Hierarchical bayesian language models for conversational speech recognition. *IEEE Transactions on Audio, Speech, and Language Processing*, 18(8):1941–1954, November 2010.

[175] X. Huang, A. Acero, C. Chelba, L. Deng, J. Droppo, D. Duchene, J. Goodman, and H. Hon. Mipad: A multimodal interaction prototype. In *Proceedings of International Conference on Acoustics Speech and Signal Processing (ICASSP)*. 2001.

[176] Y. Huang, D. Yu, Y. Gong, and C. Liu. Semi-supervised GMM and DNN acoustic model training with multi-system combination and confidence re-calibration. In *Proceedings of Interspeech*, pages 2360–2364. 2013.

[177] E. Humphrey and J. Bello. Rethinking automatic chord recognition with convolutional neural networks. In *Proceedings of International Conference on Machine Learning and Application (ICMLA)*. 2012a.

[178] E. Humphrey, J. Bello, and Y. LeCun. Moving beyond feature design: Deep architectures and automatic feature learning in music informatics. In *Proceedings of International Symposium on Music Information Retrieval (ISMIR)*. 2012.

[179] E. Humphrey, J. Bello, and Y. LeCun. Feature learning and deep architectures: New directions for music informatics. *Journal of Intelligent Information Systems*, 2013.

[180] B. Hutchinson, L. Deng, and D. Yu. A deep architecture with bilinear modeling of hidden representations: Applications to phonetic recognition. In *Proceedings of International Conference on Acoustics Speech and Signal Processing (ICASSP)*. 2012.

[181] B. Hutchinson, L. Deng, and D. Yu. Tensor deep stacking networks. *IEEE Transactions on Pattern Analysis and Machine Intelligence*, 35:1944–1957, 2013.

[182] D. Imseng, P. Motlicek, P. Garner, and H. Bourlard. Impact of deep MLP architecture on different modeling techniques for under-resourced speech recognition. In *Proceedings of the Automatic Speech Recognition and Understanding Workshop (ASRU)*. 2013.

[183] N. Jaitly and G. Hinton. Learning a better representation of speech sound waves using restricted boltzmann machines. In *Proceedings of International Conference on Acoustics Speech and Signal Processing (ICASSP)*. 2011.

[184] N. Jaitly, P. Nguyen, and V. Vanhoucke. Application of pre-trained deep neural networks to large vocabulary speech recognition. In *Proceedings of Interspeech*. 2012.

[185] K. Jarrett, K. Kavukcuoglu, and Y. LeCun. What is the best multistage architecture for object recognition? In *Proceedings of International Conference on Computer Vision*, pages 2146–2153. 2009.

[186] H. Jiang and X. Li. Parameter estimation of statistical models using convex optimization: An advanced method of discriminative training for speech and language processing. *IEEE Signal Processing Magazine*, 27(3):115–127, 2010.

[187] B. Juang, S. Levinson, and M. Sondhi. Maximum likelihood estimation for multivariate mixture observations of Markov chains. *IEEE Transactions on Information Theory*, 32:307–309, 1986.

[188] B.-H. Juang, W. Chou, and C.-H. Lee. Minimum classification error rate methods for speech recognition. *IEEE Transactions On Speech and Audio Processing*, 5:257–265, 1997.

[189] S. Kahou et al. Combining modality specific deep neural networks for emotion recognition in video. In *Proceedings of International Conference on Multimodal Interaction (ICMI)*. 2013.

[190] S. Kang, X. Qian, and H. Meng. Multi-distribution deep belief network for speech synthesis. In *Proceedings of International Conference on Acoustics Speech and Signal Processing (ICASSP)*, pages 8012–8016. 2013.

[191] Y. Kashiwagi, D. Saito, N. Minematsu, and K. Hirose. Discriminative piecewise linear transformation based on deep learning for noise robust automatic speech recognition. In *Proceedings of the Automatic Speech Recognition and Understanding Workshop (ASRU)*. 2013.

[192] K. Kavukcuoglu, P. Sermanet, Y. Boureau, K. Gregor, M. Mathieu, and Y. LeCun. Learning convolutional feature hierarchies for visual recognition. In *Proceedings of Neural Information Processing Systems (NIPS)*. 2010.

[193] H. Ketabdar and H. Bourlard. Enhanced phone posteriors for improving speech recognition systems. *IEEE Transactions on Audio, Speech, and Language Processing*, 18(6):1094–1106, August 2010.

[194] B. Kingsbury. Lattice-based optimization of sequence classification criteria for neural-network acoustic modeling. In *Proceedings of International Conference on Acoustics Speech and Signal Processing (ICASSP)*. 2009.

[195] B. Kingsbury, T. Sainath, and H. Soltau. Scalable minimum bayes risk training of deep neural network acoustic models using distributed hessian-free optimization. In *Proceedings of Interspeech*. 2012.

[196] R. Kiros, R. Zemel, and R. Salakhutdinov. Multimodal neural language models. In *Proceedings of Neural Information Processing Systems (NIPS) Deep Learning Workshop*. 2013.

[197] T. Ko and B. Mak. Eigentriphones for context-dependent acoustic modeling. *IEEE Transactions on Audio, Speech, and Language Processing*, 21(6):1285–1294, 2013.

[198] A. Krizhevsky, I. Sutskever, and G. Hinton. Imagenet classification with deep convolutional neural networks. In *Proceedings of Neural Information Processing Systems (NIPS)*. 2012.

[199] Y. Kubo, T. Hori, and A. Nakamura. Integrating deep neural networks into structural classification approach based on weighted finite-state transducers. In *Proceedings of Interspeech*. 2012.

[200] R. Kurzweil. *How to Create a Mind*. Viking Books, December 2012.

[201] P. Lal and S. King. Cross-lingual automatic speech recognition using tandem features. *IEEE Transactions on Audio, Speech, and Language Processing*, 21(12):2506–2515, December 2013.

[202] K. Lang, A. Waibel, and G. Hinton. A time-delay neural network architecture for isolated word recognition. *Neural Networks*, 3(1):23–43, 1990.

[203] H. Larochelle and Y. Bengio. Classification using discriminative restricted boltzmann machines. In *Proceedings of International Conference on Machine Learning (ICML)*. 2008.

[204] D. Le and P. Mower. Emotion recognition from spontaneous speech using hidden markov models with deep belief networks. In *Proceedings of the Automatic Speech Recognition and Understanding Workshop (ASRU)*. 2013.

[205] H. Le, A. Allauzen, G. Wisniewski, and F. Yvon. Training continuous space language models: Some practical issues. In *Proceedings of Empirical Methods in Natural Language Processing (EMNLP)*, pages 778–788. 2010.

[206] H. Le, I. Oparin, A. Allauzen, J. Gauvain, and F. Yvon. Structured output layer neural network language model. In *Proceedings of International Conference on Acoustics Speech and Signal Processing (ICASSP)*. 2011.

[207] H. Le, I. Oparin, A. Allauzen, J.-L. Gauvain, and F. Yvon. Structured output layer neural network language models for speech recognition. *IEEE Transactions on Audio, Speech, and Language Processing*, 21(1):197–206, January 2013.

[208] Q. Le, J. Ngiam, A. Coates, A. Lahiri, B. Prochnow, and A. Ng. On optimization methods for deep learning. In *Proceedings of International Conference on Machine Learning (ICML)*. 2011.

[209] Q. Le, M. Ranzato, R. Monga, M. Devin, G. Corrado, K. Chen, J. Dean, and A. Ng. Building high-level features using large scale unsupervised learning. In *Proceedings of International Conference on Machine Learning (ICML)*. 2012.

[210] Y. LeCun. Learning invariant feature hierarchies. In *Proceedings of European Conference on Computer Vision (ECCV)*. 2012.

[211] Y. LeCun and Y. Bengio. Convolutional networks for images, speech, and time series. In M. Arbib, editor, *The Handbook of Brain Theory and Neural Networks*, pages 255–258. MIT Press, Cambridge, Massachusetts, 1995.

[212] Y. LeCun, L. Bottou, Y. Bengio, and P. Haffner. Gradient-based learning applied to document recognition. *Proceedings of the IEEE*, 86:2278–2324, 1998.

[213] Y. LeCun, S. Chopra, M. Ranzato, and F. Huang. Energy-based models in document recognition and computer vision. In *Proceedings of International Conference on Document Analysis and Recognition (ICDAR)*. 2007.

[214] C.-H. Lee. From knowledge-ignorant to knowledge-rich modeling: A new speech research paradigm for next-generation automatic speech recognition. In *Proceedings of International Conference on Spoken Language Processing (ICSLP)*, pages 109–111. 2004.

[215] H. Lee, R. Grosse, R. Ranganath, and A. Ng. Convolutional deep belief networks for scalable unsupervised learning of hierarchical representations. In *Proceedings of International Conference on Machine Learning (ICML)*. 2009.

[216] H. Lee, R. Grosse, R. Ranganath, and A. Ng. Unsupervised learning of hierarchical representations with convolutional deep belief networks. *Communications of the Association for Computing Machinery (ACM)*, 54(10):95–103, October 2011.

[217] H. Lee, Y. Largman, P. Pham, and A. Ng. Unsupervised feature learning for audio classification using convolutional deep belief networks. In *Proceedings of Neural Information Processing Systems (NIPS)*. 2010.

[218] P. Lena, K. Nagata, and P. Baldi. Deep spatiotemporal architectures and learning for protein structure prediction. In *Proceedings of Neural Information Processing Systems (NIPS)*. 2012.

[219] S. Levine. Exploring deep and recurrent architectures for optimal control. arXiv:1311.1761v1.

[220] J. Li, L. Deng, Y. Gong, and R. Haeb-Umbach. An overview of noise-robust automatic speech recognition. *IEEE/Association for Computing Machinery (ACM) Transactions on Audio, Speech, and Language Processing*, pages 1–33, 2014.

[221] J. Li, D. Yu, J. Huang, and Y. Gong. Improving wideband speech recognition using mixed-bandwidth training data in CD-DNN-HMM. In *Proceedings of IEEE Spoken Language Technology (SLT)*. 2012.

[222] L. Li, Y. Zhao, D. Jiang, and Y. Zhang etc. Hybrid deep neural network–hidden markov model (DNN-HMM) based speech emotion recognition. In *Proceedings Conference on Affective Computing and Intelligent Interaction (ACII)*, pages 312–317. September 2013.

[223] H. Liao. Speaker adaptation of context dependent deep neural networks. In *Proceedings of International Conference on Acoustics Speech and Signal Processing (ICASSP)*. 2013.

[224] H. Liao, E. McDermott, and A. Senior. Large scale deep neural network acoustic modeling with semi-supervised training data for youtube video transcription. In *Proceedings of the Automatic Speech Recognition and Understanding Workshop (ASRU)*. 2013.

[225] H. Lin, L. Deng, D. Yu, Y. Gong, A. Acero, and C.-H. Lee. A study on multilingual acoustic modeling for large vocabulary ASR. In *Proceedings of International Conference on Acoustics Speech and Signal Processing (ICASSP)*. 2009.

[226] Y. Lin, F. Lv, S. Zhu, M. Yang, T. Cour, K. Yu, L. Cao, and T. Huang. Large-scale image classification: Fast feature extraction and SVM training. In *Proceedings of Computer Vision and Pattern Recognition (CVPR)*. 2011.

[227] Z. Ling, L. Deng, and D. Yu. Modeling spectral envelopes using restricted boltzmann machines and deep belief networks for statistical parametric speech synthesis. *IEEE Transactions on Audio Speech Language Processing*, 21(10):2129–2139, 2013.

[228] Z. Ling, L. Deng, and D. Yu. Modeling spectral envelopes using restricted boltzmann machines for statistical parametric speech synthesis. In *International Conference on Acoustics Speech and Signal Processing (ICASSP)*, pages 7825–7829. 2013.

[229] Z. Ling, K. Richmond, and J. Yamagishi. Articulatory control of HMM-based parametric speech synthesis using feature-space-switched multiple regression. *IEEE Transactions on Audio, Speech, and Language Processing*, 21, January 2013.

[230] L. Lu, K. Chin, A. Ghoshal, and S. Renals. Joint uncertainty decoding for noise robust subspace gaussian mixture models. *IEEE Transactions on Audio, Speech, and Language Processing*, 21(9):1791–1804, 2013.

[231] J. Ma and L. Deng. A path-stack algorithm for optimizing dynamic regimes in a statistical hidden dynamical model of speech. *Computer, Speech and Language*, 2000.

[232] J. Ma and L. Deng. Efficient decoding strategies for conversational speech recognition using a constrained nonlinear state-space model. *IEEE Transactions on Speech and Audio Processing*, 11(6):590–602, 2003.

[233] J. Ma and L. Deng. Target-directed mixture dynamic models for spontaneous speech recognition. *IEEE Transactions on Speech and Audio Processing*, 12(1):47–58, 2004.

[234] A. Maas, A. Hannun, and A. Ng. Rectifier nonlinearities improve neural network acoustic models. *International Conference on Machine Learning (ICML) Workshop on Deep Learning for Audio, Speech, and Language Processing*, 2013.

[235] A. Maas, Q. Le, T. O'Neil, O. Vinyals, P. Nguyen, and P. Ng. Recurrent neural networks for noise reduction in robust ASR. In *Proceedings of Interspeech*. 2012.

[236] C. Manning, P. Raghavan, and H. Schütze. *Introduction to Information Retrieval*. Cambridge University Press, 2009.

[237] J. Markoff. Scientists see promise in deep-learning programs. *New York Times*, November 24 2012.

[238] J. Martens. Deep learning with hessian-free optimization. In *Proceedings of International Conference on Machine Learning (ICML)*. 2010.

[239] J. Martens and I. Sutskever. Learning recurrent neural networks with hessian-free optimization. In *Proceedings of International Conference on Machine Learning (ICML)*. 2011.

[240] D. McAllester. A PAC-bayesian tutorial with a dropout bound. ArXive1307.2118, July 2013.

[241] I. McGraw, I. Badr, and J. R. Glass. Learning lexicons from speech using a pronunciation mixture model. *IEEE Transactions on Audio, Speech, and Language Processing*, 21(2):357,366, February 2013.

[242] G. Mesnil, X. He, L. Deng, and Y. Bengio. Investigation of recurrent-neural-network architectures and learning methods for spoken language understanding. In *Proceedings of Interspeech*. 2013.

[243] Y. Miao and F. Metze. Improving low-resource CD-DNN-HMM using dropout and multilingual DNN training. In *Proceedings of Interspeech*. 2013.

[244] Y. Miao, S. Rawat, and F. Metze. Deep maxout networks for low resource speech recognition. In *Proceedings of the Automatic Speech Recognition and Understanding Workshop (ASRU)*. 2013.

[245] T. Mikolov. Statistical language models based on neural networks. Ph.D. thesis, Brno University of Technology, 2012.

[246] T. Mikolov, K. Chen, G. Corrado, and J. Dean. Efficient estimation of word representations in vector space. In *Proceedings of International Conference on Learning Representations (ICLR)*. 2013.

[247] T. Mikolov, A. Deoras, D. Povey, L. Burget, and J. Cernocky. Strategies for training large scale neural network language models. In *Proceedings of the IEEE Automatic Speech Recognition and Understanding Workshop (ASRU)*. 2011.

[248] T. Mikolov, M. Karafiat, L. Burget, J. Cernocky, and S. Khudanpur. Recurrent neural network based language model. In *Proceedings of International Conference on Acoustics Speech and Signal Processing (ICASSP)*, pages 1045–1048. 2010.

[249] T. Mikolov, Q. Le, and I. Sutskever. Exploiting similarities among languages for machine translation. arXiv:1309.4168v1, 2013.

[250] T. Mikolov, I. Sutskever, K. Chen, G. Corrado, and J. Dean. Distributed representations of words and phrases and their compositionality. In *Proceedings of Neural Information Processing Systems (NIPS)*. 2013.

[251] Y. Minami, E. McDermott, A. Nakamura, and S. Katagiri. A recognition method with parametric trajectory synthesized using direct relations between static and dynamic feature vector time series. In *Proceedings of International Conference on Acoustics Speech and Signal Processing (ICASSP)*, pages 957–960. 2002.

[252] A. Mnih and G. Hinton. Three new graphical models for statistical language modeling. In *Proceedings of International Conference on Machine Learning (ICML)*, pages 641–648. 2007.

[253] A. Mnih and G. Hinton. A scalable hierarchical distributed language model. In *Proceedings of Neural Information Processing Systems (NIPS)*, pages 1081–1088. 2008.

[254] A. Mnih and K. Kavukcuoglu. Learning word embeddings efficiently with noise-contrastive estimation. In *Proceedings of Neural Information Processing Systems (NIPS)*. 2013.

[255] A. Mnih and W.-T. Teh. A fast and simple algorithm for training neural probabilistic language models. In *Proceedings of International Conference on Machine Learning (ICML)*, pages 1751–1758. 2012.

[256] V. Mnih, K. Kavukcuoglu, D. Silver, A. Graves, I. Antonoglou, D. Wierstra, and M. Riedmiller. Playing arari with deep reinforcement learning. *Neural Information Processing Systems (NIPS) Deep Learning Workshop*, 2013. also arXiv:1312.5602v1.

[257] A. Mohamed, G. Dahl, and G. Hinton. Deep belief networks for phone recognition. In *Proceedings of Neural Information Processing Systems (NIPS) Workshop Deep Learning for Speech Recognition and Related Applications*. 2009.

[258] A. Mohamed, G. Dahl, and G. Hinton. Acoustic modeling using deep belief networks. *IEEE Transactions on Audio, Speech, & Language Processing*, 20(1), January 2012.

[259] A. Mohamed, G. Hinton, and G. Penn. Understanding how deep belief networks perform acoustic modelling. In *Proceedings of International Conference on Acoustics Speech and Signal Processing (ICASSP)*. 2012.

[260] A. Mohamed, D. Yu, and L. Deng. Investigation of full-sequence training of deep belief networks for speech recognition. In *Proceedings of Interspeech*. 2010.

[261] N. Morgan. Deep and wide: Multiple layers in automatic speech recognition. *IEEE Transactions on Audio, Speech, & Language Processing*, 20(1), January 2012.

[262] N. Morgan, Q. Zhu, A. Stolcke, K. Sonmez, S. Sivadas, T. Shinozaki, M. Ostendorf, P. Jain, H. Hermansky, D. Ellis, G. Doddington, B. Chen, O. Cretin, H. Bourlard, and M. Athineos. Pushing the envelope — aside [speech recognition]. *IEEE Signal Processing Magazine*, 22(5):81–88, September 2005.

[263] F. Morin and Y. Bengio. Hierarchical probabilistic neural network language models. In *Proceedings of Artificial Intelligence and Statistics (AISTATS)*. 2005.

[264] K. Murphy. *Machine Learning — A Probabilistic Perspective*. The MIT Press, 2012.

[265] V. Nair and G. Hinton. 3-d object recognition with deep belief nets. In *Proceedings of Neural Information Processing Systems (NIPS)*. 2009.

[266] T. Nakashika, R. Takashima, T. Takiguchi, and Y. Ariki. Voice conversion in high-order eigen space using deep belief nets. In *Proceedings of Interspeech*. 2013.

[267] H. Ney. Speech translation: Coupling of recognition and translation. In *Proceedings of International Conference on Acoustics Speech and Signal Processing (ICASSP)*. 1999.

[268] J. Ngiam, Z. Chen, P. Koh, and A. Ng. Learning deep energy models. In *Proceedings of International Conference on Machine Learning (ICML)*. 2011.

[269] J. Ngiam, A. Khosla, M. Kim, J. Nam, H. Lee, and A. Ng. Multimodal deep learning. In *Proceedings of International Conference on Machine Learning (ICML)*. 2011.

[270] M. Norouzi, T. Mikolov, S. Bengio, J. Shlens, A. Frome, G. Corrado, and J. Dean. Zero-shot learning by convex combination of semantic embeddings. arXiv:1312.5650v2, 2013.

[271] N. Oliver, A. Garg, and E. Horvitz. Layered representations for learning and inferring office activity from multiple sensory channels. *Computer Vision and Image Understanding*, 96:163–180, 2004.

[272] B. Olshausen. Can 'deep learning' offer deep insights about visual representation? *Neural Information Processing Systems (NIPS) Workshop on Deep Learning and Unsupervised Feature Learning*, 2012.

[273] M. Ostendorf. Moving beyond the 'beads-on-a-string' model of speech. In *Proceedings of the Automatic Speech Recognition and Understanding Workshop (ASRU)*. 1999.

[274] M. Ostendorf, V. Digalakis, and O. Kimball. From HMMs to segment models: A unified view of stochastic modeling for speech recognition. *IEEE Transactions on Speech and Audio Processing*, 4(5), September 1996.

[275] L. Oudre, C. Fevotte, and Y. Grenier. Probabilistic template-based chord recognition. *IEEE Transactions on Audio, Speech, and Language Processing*, 19(8):2249–2259, November 2011.

[276] H. Palangi, L. Deng, and R. Ward. Learning input and recurrent weight matrices in echo state networks. *Neural Information Processing Systems (NIPS) Deep Learning Workshop*, December 2013.

[277] H. Palangi, R. Ward, and L. Deng. Using deep stacking network to improve structured compressive sensing with multiple measurement vectors. In *Proceedings of International Conference on Acoustics Speech and Signal Processing (ICASSP)*. 2013.

[278] G. Papandreou, A. Katsamanis, V. Pitsikalis, and P. Maragos. Adaptive multimodal fusion by uncertainty compensation with application to audiovisual speech recognition. *IEEE Transactions on Audio, Speech, and Language Processing*, 17:423–435, 2009.

[279] R. Pascanu, C. Gulcehre, K. Cho, and Y. Bengio. How to construct deep recurrent neural networks. In *Proceedings of International Conference on Learning Representations (ICLR)*. 2014.

[280] R. Pascanu, T. Mikolov, and Y. Bengio. On the difficulty of training recurrent neural networks. In *Proceedings of International Conference on Machine Learning (ICML)*. 2013.

[281] J. Peng, L. Bo, and J. Xu. Conditional neural fields. In *Proceedings of Neural Information Processing Systems (NIPS)*. 2009.

[282] P. Picone, S. Pike, R. Regan, T. Kamm, J. bridle, L. Deng, Z. Ma, H. Richards, and M. Schuster. Initial evaluation of hidden dynamic models on conversational speech. In *Proceedings of International Conference on Acoustics Speech and Signal Processing (ICASSP)*. 1999.

[283] J. Pinto, S. Garimella, M. Magimai-Doss, H. Hermansky, and H. Bourlard. Analysis of MLP-based hierarchical phone posterior probability estimators. *IEEE Transactions on Audio, Speech, and Language Processing*, 19(2), February 2011.

[284] C. Plahl, T. Sainath, B. Ramabhadran, and D. Nahamoo. Improved pre-training of deep belief networks using sparse encoding symmetric machines. In *Proceedings of International Conference on Acoustics Speech and Signal Processing (ICASSP)*. 2012.

[285] C. Plahl, R. Schlüter, and H. Ney. Hierarchical bottleneck features for LVCSR. In *Proceedings of Interspeech*. 2010.

[286] T. Plate. Holographic reduced representations. *IEEE Transactions on Neural Networks*, 6(3):623–641, May 1995.

[287] T. Poggio. How the brain might work: The role of information and learning in understanding and replicating intelligence. In G. Jacovitt, A. Pettorossi, R. Consolo, and V. Senni, editors, *Information: Science and Technology for the New Century*, pages 45–61. Lateran University Press, 2007.

[288] J. Pollack. Recursive distributed representations. *Artificial Intelligence*, 46:77–105, 1990.

[289] H. Poon and P. Domingos. Sum-product networks: A new deep architecture. In *Proceedings of Uncertainty in Artificial Intelligence*. 2011.

[290] D. Povey and P. Woodland. Minimum phone error and I-smoothing for improved discriminative training. In *Proceedings of International Conference on Acoustics Speech and Signal Processing (ICASSP)*. 2002.

[291] R. Prabhavalkar and E. Fosler-Lussier. Backpropagation training for multilayer conditional random field based phone recognition. In *Proceedings of International Conference on Acoustics Speech and Signal Processing (ICASSP)*. 2010.

[292] A. Prince and P. Smolensky. Optimality: From neural networks to universal grammar. *Science*, 275:1604–1610, 1997.

[293] L. Rabiner. A tutorial on hidden markov models and selected applications in speech recognition. In *Proceedings of the IEEE*, pages 257–286. 1989.

[294] M. Ranzato, Y. Boureau, and Y. LeCun. Sparse feature learning for deep belief networks. In *Proceedings of Neural Information Processing Systems (NIPS)*. 2007.

[295] M. Ranzato, S. Chopra, Y. LeCun, and F.-J. Huang. Energy-based models in document recognition and computer vision. In *Proceedings of International Conference on Document Analysis and Recognition (ICDAR)*. 2007.

[296] M. Ranzato and G. Hinton. Modeling pixel means and covariances using factorized third-order boltzmann machines. In *Proceedings of Computer Vision and Pattern Recognition (CVPR)*. 2010.

[297] M. Ranzato, C. Poultney, S. Chopra, and Y. LeCun. Efficient learning of sparse representations with an energy-based model. In *Proceedings of Neural Information Processing Systems (NIPS)*. 2006.

[298] M. Ranzato, J. Susskind, V. Mnih, and G. Hinton. On deep generative models with applications to recognition. In *Proceedings of Computer Vision and Pattern Recognition (CVPR)*. 2011.

[299] C. Rathinavalu and L. Deng. Construction of state-dependent dynamic parameters by maximum likelihood: Applications to speech recognition. *Signal Processing*, 55(2):149–165, 1997.

[300] S. Rennie, K. Fouset, and P. Dognin. Factorial hidden restricted boltzmann machines for noise robust speech recognition. In *Proceedings of International Conference on Acoustics Speech and Signal Processing (ICASSP)*. 2012.

[301] S. Rennie, H. Hershey, and P. Olsen. Single-channel multi-talker speech recognition — graphical modeling approaches. *IEEE Signal Processing Magazine*, 33:66–80, 2010.

[302] M. Riedmiller and H. Braun. A direct adaptive method for faster backpropagation learning: The RPROP algorithm. In *Proceedings of the IEEE International Conference on Neural Networks*. 1993.

[303] S. Rifai, P. Vincent, X. Muller, X. Glorot, and Y. Bengio. Contractive autoencoders: Explicit invariance during feature extraction. In *Proceedings of International Conference on Machine Learning (ICML)*, pages 833–840. 2011.

[304] A. Robinson. An application of recurrent nets to phone probability estimation. *IEEE Transactions on Neural Networks*, 5:298–305, 1994.

[305] T. Sainath, L. Horesh, B. Kingsbury, A. Aravkin, and B. Ramabhadran. Accelerating hessian-free optimization for deep neural networks by implicit pre-conditioning and sampling. arXiv: 1309.1508v3, 2013.

[306] T. Sainath, B. Kingsbury, A. Mohamed, G. Dahl, G. Saon, H. Soltau, T. Beran, A. Aravkin, and B. Ramabhadran. Improvements to deep convolutional neural networks for LVCSR. In *Proceedings of the Automatic Speech Recognition and Understanding Workshop (ASRU)*. 2013.

[307] T. Sainath, B. Kingsbury, A. Mohamed, and B. Ramabhadran. Learning filter banks within a deep neural network framework. In *Proceedings of The Automatic Speech Recognition and Understanding Workshop (ASRU)*. 2013.

[308] T. Sainath, B. Kingsbury, and B. Ramabhadran. Autoencoder bottleneck features using deep belief networks. In *Proceedings of International Conference on Acoustics Speech and Signal Processing (ICASSP)*. 2012.

[309] T. Sainath, B. Kingsbury, B. Ramabhadran, P. Novak, and A. Mohamed. Making deep belief networks effective for large vocabulary continuous speech recognition. In *Proceedings of the Automatic Speech Recognition and Understanding Workshop (ASRU)*. 2011.

[310] T. Sainath, B. Kingsbury, V. Sindhwani, E. Arisoy, and B. Ramabhadran. Low-rank matrix factorization for deep neural network training with high-dimensional output targets. In *Proceedings of International Conference on Acoustics Speech and Signal Processing (ICASSP)*. 2013.

[311] T. Sainath, B. Kingsbury, H. Soltau, and B. Ramabhadran. Optimization techniques to improve training speed of deep neural networks for large speech tasks. *IEEE Transactions on Audio, Speech, and Language Processing*, 21(11):2267–2276, November 2013.

[312] T. Sainath, A. Mohamed, B. Kingsbury, and B. Ramabhadran. Convolutional neural networks for LVCSR. In *Proceedings of International Conference on Acoustics Speech and Signal Processing (ICASSP)*. 2013.

[313] T. Sainath, B. Ramabhadran, M. Picheny, D. Nahamoo, and D. Kanevsky. Exemplar-based sparse representation features: From TIMIT to LVCSR. *IEEE Transactions on Speech and Audio Processing*, November 2011.

[314] R. Salakhutdinov and G. Hinton. Semantic hashing. In *Proceedings of Special Interest Group on Information Retrieval (SIGIR) Workshop on Information Retrieval and Applications of Graphical Models*. 2007.

[315] R. Salakhutdinov and G. Hinton. Deep boltzmann machines. In *Proceedings of Artificial Intelligence and Statistics (AISTATS)*. 2009.

[316] R. Salakhutdinov and G. Hinton. A better way to pretrain deep boltzmann machines. In *Proceedings of Neural Information Processing Systems (NIPS)*. 2012.

[317] G. Saon, H. Soltau, D. Nahamoo, and M. Picheny. Speaker adaptation of neural network acoustic models using i-vectors. In *Proceedings of the Automatic Speech Recognition and Understanding Workshop (ASRU)*. 2013.

[318] R. Sarikaya, G. Hinton, and B. Ramabhadran. Deep belief nets for natural language call-routing. In *Proceedings of International Conference on Acoustics Speech and Signal Processing (ICASSP)*, pages 5680–5683. 2011.

[319] E. Schmidt and Y. Kim. Learning emotion-based acoustic features with deep belief networks. In *Proceedings IEEE of Signal Processing to Audio and Acoustics*. 2011.

[320] H. Schwenk. Continuous space translation models for phrase-based statistical machine translation. In *Proceedings of Computional Linguistics*. 2012.

[321] H. Schwenk, A. Rousseau, and A. Mohammed. Large, pruned or continuous space language models on a gpu for statistical machine translation. In *Proceedings of the Joint Human Language Technology Conference and the North American Chapter of the Association of Computational Linguistics (HLT-NAACL) 2012 Workshop on the future of language modeling for Human Language Technology (HLT)*, pages 11–19.

[322] F. Seide, H. Fu, J. Droppo, G. Li, and D. Yu. On parallelizability of stochastic gradient descent for speech DNNs. In *Proceedings of International Conference on Acoustics Speech and Signal Processing (ICASSP)*. 2014.

[323] F. Seide, G. Li, X. Chen, and D. Yu. Feature engineering in context-dependent deep neural networks for conversational speech transcription. In *Proceedings of the Automatic Speech Recognition and Understanding Workshop (ASRU)*, pages 24–29. 2011.

[324] F. Seide, G. Li, and D. Yu. Conversational speech transcription using context-dependent deep neural networks. In *Proceedings of Interspeech*, pages 437–440. 2011.

[325] M. Seltzer, D. Yu, and E. Wang. An investigation of deep neural networks for noise robust speech recognition. In *Proceedings of International Conference on Acoustics Speech and Signal Processing (ICASSP)*. 2013.

[326] M. Shannon, H. Zen, and W. Byrne. Autoregressive models for statistical parametric speech synthesis. *IEEE Transactions on Audio, Speech, Language Processing*, 21(3):587–597, 2013.

[327] H. Sheikhzadeh and L. Deng. Waveform-based speech recognition using hidden filter models: Parameter selection and sensitivity to power normalization. *IEEE Transactions on on Speech and Audio Processing (ICASSP)*, 2:80–91, 1994.

[328] Y. Shen, X. He, J. Gao, L. Deng, and G. Mesnil. Learning semantic representations using convolutional neural networks for web search. In *Proceedings World Wide Web*. 2014.

[329] K. Simonyan, A. Vedaldi, and A. Zisserman. Deep fisher networks for large-scale image classification. In *Proceedings of Neural Information Processing Systems (NIPS)*. 2013.

[330] M. Siniscalchi, J. Li, and C. Lee. Hermitian polynomial for speaker adaptation of connectionist speech recognition systems. *IEEE Transactions on Audio, Speech, and Language Processing*, 21(10):2152–2161, 2013a.

[331] M. Siniscalchi, T. Svendsen, and C.-H. Lee. A bottom-up modular search approach to large vocabulary continuous speech recognition. *IEEE Transactions on Audio, Speech, Language Processing*, 21, 2013.

[332] M. Siniscalchi, D. Yu, L. Deng, and C.-H. Lee. Exploiting deep neural networks for detection-based speech recognition. *Neurocomputing*, 106:148–157, 2013.

[333] M. Siniscalchi, D. Yu, L. Deng, and C.-H. Lee. Speech recognition using long-span temporal patterns in a deep network model. *IEEE Signal Processing Letters*, 20(3):201–204, March 2013.

[334] G. Sivaram and H. Hermansky. Sparse multilayer perceptrons for phoneme recognition. *IEEE Transactions on Audio, Speech, & Language Processing*, 20(1), January 2012.

[335] P. Smolensky. Tensor product variable binding and the representation of symbolic structures in connectionist systems. *Artificial Intelligence*, 46:159–216, 1990.

[336] P. Smolensky and G. Legendre. *The Harmonic Mind — From Neural Computation to Optimality-Theoretic Grammar*. The MIT Press, Cambridge, MA, 2006.

[337] J. Snoek, H. Larochelle, and R. Adams. Practical bayesian optimization of machine learning algorithms. In *Proceedings of Neural Information Processing Systems (NIPS)*. 2012.

[338] R. Socher. New directions in deep learning: Structured models, tasks, and datasets. *Neural Information Processing Systems (NIPS) Workshop on Deep Learning and Unsupervised Feature Learning*, 2012.

[339] R. Socher, Y. Bengio, and C. Manning. Deep learning for NLP. *Tutorial at Association of Computational Logistics (ACL), 2012, and North American Chapter of the Association of Computational Linguistics (NAACL)*, 2013. http://www.socher.org/index.php/DeepLearning Tutorial.

[340] R. Socher, D. Chen, C. Manning, and A. Ng. Reasoning with neural tensor networks for knowledge base completion. In *Proceedings of Neural Information Processing Systems (NIPS)*. 2013.

[341] R. Socher and L. Fei-Fei. Connecting modalities: Semi-supervised segmentation and annotation of images using unaligned text corpora. In *Proceedings of Computer Vision and Pattern Recognition (CVPR)*. 2010.

[342] R. Socher, M. Ganjoo, H. Sridhar, O. Bastani, C. Manning, and A. Ng. Zero-shot learning through cross-modal transfer. In *Proceedings of Neural Information Processing Systems (NIPS)*. 2013b.

[343] R. Socher, Q. Le, C. Manning, and A. Ng. Grounded compositional semantics for finding and describing images with sentences. *Neural Information Processing Systems (NIPS) Deep Learning Workshop*, 2013c.

[344] R. Socher, C. Lin, A. Ng, and C. Manning. Parsing natural scenes and natural language with recursive neural networks. In *Proceedings of International Conference on Machine Learning (ICML)*. 2011.

[345] R. Socher, J. Pennington, E. Huang, A. Ng, and C. Manning. Dynamic pooling and unfolding recursive autoencoders for paraphrase detection. In *Proceedings of Neural Information Processing Systems (NIPS)*. 2011.

[346] R. Socher, J. Pennington, E. Huang, A. Ng, and C. Manning. Semi-supervised recursive autoencoders for predicting sentiment distributions. In *Proceedings of Empirical Methods in Natural Language Processing (EMNLP)*. 2011.

[347] R. Socher, A. Perelygin, J. Wu, J. Chuang, C. Manning, A. Ng, and C. Potts. Recursive deep models for semantic compositionality over a sentiment treebank. In *Proceedings of Empirical Methods in Natural Language Processing (EMNLP)*. 2013.

[348] N. Srivastava and R. Salakhutdinov. Multimodal learning with deep boltzmann machines. In *Proceedings of Neural Information Processing Systems (NIPS)*. 2012.

[349] N. Srivastava and R. Salakhutdinov. Discriminative transfer learning with tree-based priors. In *Proceedings of Neural Information Processing Systems (NIPS)*. 2013.

[350] R. Srivastava, J. Masci, S. Kazerounian, F. Gomez, and J. Schmidhuber. Compete to compute. In *Proceedings of Neural Information Processing Systems (NIPS)*. 2013.

[351] T. Stafylakis, P. Kenny, M. Senoussaoui, and P. Dumouchel. Preliminary investigation of boltzmann machine classifiers for speaker recognition. In *Proceedings of Odyssey*, pages 109–116. 2012.

[352] V. Stoyanov, A. Ropson, and J. Eisner. Empirical risk minimization of graphical model parameters given approximate inference, decoding, and model structure. In *Proceedings of Artificial Intelligence and Statistics (AISTATS)*. 2011.

[353] H. Su, G. Li, D. Yu, and F. Seide. Error back propagation for sequence training of context-dependent deep networks for conversational speech transcription. In *Proceedings of International Conference on Acoustics Speech and Signal Processing (ICASSP)*. 2013.

[354] A. Subramanya, L. Deng, Z. Liu, and Z. Zhang. Multi-sensory speech processing: Incorporating automatically extracted hidden dynamic information. In *Proceedings of IEEE International Conference on Multimedia & Expo (ICME)*. Amsterdam, July 2005.

[355] J. Sun and L. Deng. An overlapping-feature based phonological model incorporating linguistic constraints: Applications to speech recognition. *Journal on Acoustical Society of America*, 111(2):1086–1101, 2002.

[356] I. Sutskever. Training recurrent neural networks. Ph.D. Thesis, University of Toronto, 2013.

[357] I. Sutskever, J. Martens, and G. Hinton. Generating text with recurrent neural networks. In *Proceedings of International Conference on Machine Learning (ICML)*. 2011.

[358] Y. Tang and C. Eliasmith. Deep networks for robust visual recognition. In *Proceedings of International Conference on Machine Learning (ICML)*. 2010.

[359] Y. Tang and R. Salakhutdinov. *Learning Stochastic Feedforward Neural Networks*. NIPS, 2013.

[360] A. Tarralba, R. Fergus, and Y. Weiss. Small codes and large image databases for recognition. In *Proceedings of Computer Vision and Pattern Recognition (CVPR)*. 2008.

[361] G. Taylor, G. E. Hinton, and S. Roweis. Modeling human motion using binary latent variables. In *Proceedings of Neural Information Processing Systems (NIPS)*. 2007.

[362] S. Thomas, M. Seltzer, K. Church, and H. Hermansky. Deep neural network features and semi-supervised training for low resource speech recognition. In *Proceedings of Interspeech*. 2013.

[363] T. Tieleman. Training restricted boltzmann machines using approximations to the likelihood gradient. In *Proceedings of International Conference on Machine Learning (ICML)*. 2008.

[364] K. Tokuda, Y. Nankaku, T. Toda, H. Zen, H. Yamagishi, and K. Oura. Speech synthesis based on hidden markov models. *Proceedings of the IEEE*, 101(5):1234–1252, 2013.

[365] F. Triefenbach, A. Jalalvand, K. Demuynck, and J.-P. Martens. Acoustic modeling with hierarchical reservoirs. *IEEE Transactions on Audio, Speech, and Language Processing*, 21(11):2439–2450, November 2013.

[366] G. Tur, L. Deng, D. Hakkani-Tür, and X. He. Towards deep understanding: Deep convex networks for semantic utterance classification. In *Proceedings of International Conference on Acoustics Speech and Signal Processing (ICASSP)*. 2012.

[367] J. Turian, L. Ratinov, and Y. Bengio. Word representations: A simple and general method for semi-supervised learning. In *Proceedings of Association for Computational Linguistics (ACL)*. 2010.

[368] Z. Tüske, M. Sundermeyer, R. Schlüter, and H. Ney. Context-dependent MLPs for LVCSR: TANDEM, hybrid or both? In *Proceedings of Interspeech*. 2012.

[369] B. Uria, S. Renals, and K. Richmond. A deep neural network for acoustic-articulatory speech inversion. *Neural Information Processing Systems (NIPS) Workshop on Deep Learning and Unsupervised Feature Learning*, 2011.

[370] R. van Dalen and M. Gales. Extended VTS for noise-robust speech recognition. *IEEE Transactions on Audio, Speech, and Language Processing*, 19(4):733–743, 2011.

[371] A. van den Oord, S. Dieleman, and B. Schrauwen. Deep content-based music recommendation. In *Proceedings of Neural Information Processing Systems (NIPS)*. 2013.

[372] V. Vasilakakis, S. Cumani, and P. Laface. Speaker recognition by means of deep belief networks. In *Proceedings of Biometric Technologies in Forensic Science*. 2013.

[373] K. Vesely, A. Ghoshal, L. Burget, and D. Povey. Sequence-discriminative training of deep neural networks. In *Proceedings of Interspeech*. 2013.

[374] K. Vesely, M. Hannemann, and L. Burget. Semi-supervised training of deep neural networks. In *Proceedings of the Automatic Speech Recognition and Understanding Workshop (ASRU)*. 2013.

[375] P. Vincent. A connection between score matching and denoising autoencoder. *Neural Computation*, 23(7):1661–1674, 2011.

[376] P. Vincent, H. Larochelle, I. Lajoie, Y. Bengio, and P. Manzagol. Stacked denoising autoencoders: Learning useful representations in a deep network with a local denoising criterion. *Journal of Machine Learning Research*, 11:3371–3408, 2010.

[377] O. Vinyals, Y. Jia, L. Deng, and T. Darrell. Learning with recursive perceptual representations. In *Proceedings of Neural Information Processing Systems (NIPS)*. 2012.

[378] O. Vinyals and D. Povey. Krylov subspace descent for deep learning. In *Proceedings of Artificial Intelligence and Statistics (AISTATS)*. 2012.

[379] O. Vinyals and S. Ravuri. Comparing multilayer perceptron to deep belief network tandem features for robust ASR. In *Proceedings of International Conference on Acoustics Speech and Signal Processing (ICASSP)*. 2011.

[380] O. Vinyals, S. Ravuri, and D. Povey. Revisiting recurrent neural networks for robust ASR. In *Proceedings of International Conference on Acoustics Speech and Signal Processing (ICASSP)*. 2012.

[381] S. Wager, S. Wang, and P. Liang. Dropout training as adaptive regularization. In *Proceedings of Neural Information Processing Systems (NIPS)*. 2013.

[382] A. Waibel, T. Hanazawa, G. Hinton, K. Shikano, and K. Lang. Phoneme recognition using time-delay neural networks. *IEEE Transactions on Acoustical Speech, and Signal Processing*, 37:328–339, 1989.

[383] G. Wang and K. Sim. Context-dependent modelling of deep neural network using logistic regression. In *Proceedings of the Automatic Speech Recognition and Understanding Workshop (ASRU)*. 2013.

[384] G. Wang and K. Sim. Regression-based context-dependent modeling of deep neural networks for speech recognition. *IEEE/Association for Computing Machinery (ACM) Transactions on Audio, Speech, and Language Processing*, 2014.

[385] D. Warde-Farley, I. Goodfellow, A. Courville, and Y. Bengi. An empirical analysis of dropout in piecewise linear networks. In *Proceedings of International Conference on Learning Representations (ICLR)*. 2014.

[386] M. Welling, M. Rosen-Zvi, and G. Hinton. Exponential family harmoniums with an application to information retrieval. In *Proceedings of Neural Information Processing Systems (NIPS)*. 2005.

[387] C. Weng, D. Yu, M. Seltzer, and J. Droppo. Single-channel mixed speech recognition using deep neural networks. In *Proceedings of International Conference on Acoustics Speech and Signal Processing (ICASSP)*. 2014.

[388] J. Weston, S. Bengio, and N. Usunier. Large scale image annotation: Learning to rank with joint word-image embeddings. *Machine Learning*, 81(1):21–35, 2010.

[389] J. Weston, S. Bengio, and N. Usunier. Wsabie: Scaling up to large vocabulary image annotation. In *Proceedings of International Joint Conference on Artificial Intelligence (IJCAI)*. 2011.

[390] S. Wiesler, J. Li, and J. Xue. Investigations on hessian-free optimization for cross-entropy training of deep neural networks. In *Proceedings of Interspeech*. 2013.

[391] M. Wohlmayr, M. Stark, and F. Pernkopf. A probabilistic interaction model for multi-pitch tracking with factorial hidden markov model. *IEEE Transactions on Audio, Speech, and Language Processing*, 19(4), May 2011.

[392] D. Wolpert. Stacked generalization. *Neural Networks*, 5(2):241–259, 1992.

[393] S. J. Wright, D. Kanevsky, L. Deng, X. He, G. Heigold, and H. Li. Optimization algorithms and applications for speech and language processing. *IEEE Transactions on Audio, Speech, and Language Processing*, 21(11):2231–2243, November 2013.

[394] L. Xiao and L. Deng. A geometric perspective of large-margin training of gaussian models. *IEEE Signal Processing Magazine*, 27(6):118–123, November 2010.

[395] X. Xie and S. Seung. Equivalence of backpropagation and contrastive hebbian learning in a layered network. *Neural computation*, 15:441–454, 2003.

[396] Y. Xu, J. Du, L. Dai, and C. Lee. An experimental study on speech enhancement based on deep neural networks. *IEEE Signal Processing Letters*, 21(1):65–68, 2014.

[397] J. Xue, J. Li, and Y. Gong. Restructuring of deep neural network acoustic models with singular value decomposition. In *Proceedings of Interspeech*. 2013.

[398] S. Yamin, L. Deng, Y. Wang, and A. Acero. An integrative and discriminative technique for spoken utterance classification. *IEEE Transactions on Audio, Speech, and Language Processing*, 16:1207–1214, 2008.

[399] Z. Yan, Q. Huo, and J. Xu. A scalable approach to using DNN-derived features in GMM-HMM based acoustic modeling for LVCSR. In *Proceedings of Interspeech*. 2013.

[400] D. Yang and S. Furui. Combining a two-step CRF model and a joint source-channel model for machine transliteration. In *Proceedings of Association for Computational Linguistics (ACL)*, pages 275–280. 2010.

[401] K. Yao, D. Yu, L. Deng, and Y. Gong. A fast maximum likelihood nonlinear feature transformation method for GMM-HMM speaker adaptation. *Neurocomputing*, 2013a.

[402] K. Yao, D. Yu, F. Seide, H. Su, L. Deng, and Y. Gong. Adaptation of context-dependent deep neural networks for automatic speech recognition. In *Proceedings of International Conference on Acoustics Speech and Signal Processing (ICASSP)*. 2012.

[403] K. Yao, G. Zweig, M. Hwang, Y. Shi, and D. Yu. Recurrent neural networks for language understanding. In *Proceedings of Interspeech*. 2013.

[404] T. Yoshioka and T. Nakatani. Noise model transfer: Novel approach to robustness against nonstationary noise. *IEEE Transactions on Audio, Speech, and Language Processing*, 21(10):2182–2192, 2013.

[405] T. Yoshioka, A. Ragni, and M. Gales. Investigation of unsupervised adaptation of DNN acoustic models with filter bank input. In *Proceedings of International Conference on Acoustics Speech and Signal Processing (ICASSP)*. 2013.

[406] L. Younes. On the convergence of markovian stochastic algorithms with rapidly decreasing ergodicity rates. *Stochastics and Stochastic Reports*, 65(3):177–228, 1999.

[407] D. Yu, X. Chen, and L. Deng. Factorized deep neural networks for adaptive speech recognition. *International Workshop on Statistical Machine Learning for Speech Processing*, March 2012b.

[408] D. Yu, D. Deng, and S. Wang. Learning in the deep-structured conditional random fields. *Neural Information Processing Systems (NIPS) 2009 Workshop on Deep Learning for Speech Recognition and Related Applications*, 2009.

[409] D. Yu and L. Deng. Solving nonlinear estimation problems using splines. *IEEE Signal Processing Magazine*, 26(4):86–90, July 2009.

[410] D. Yu and L. Deng. Deep-structured hidden conditional random fields for phonetic recognition. In *Proceedings of Interspeech*. September 2010.

[411] D. Yu and L. Deng. Accelerated parallelizable neural networks learning algorithms for speech recognition. In *Proceedings of Interspeech*. 2011.

[412] D. Yu and L. Deng. Deep learning and its applications to signal and information processing. *IEEE Signal Processing Magazine*, pages 145–154, January 2011.

[413] D. Yu and L. Deng. Efficient and effective algorithms for training single-hidden-layer neural networks. *Pattern Recognition Letters*, 33:554–558, 2012.

[414] D. Yu, L. Deng, and G. E. Dahl. Roles of pre-training and fine-tuning in context-dependent DBN-HMMs for real-world speech recognition. *Neural Information Processing Systems (NIPS) 2010 Workshop on Deep Learning and Unsupervised Feature Learning*, December 2010.

[415] D. Yu, L. Deng, J. Droppo, J. Wu, Y. Gong, and A. Acero. Robust speech recognition using cepstral minimum-mean-square-error noise suppressor. *IEEE Transactions on Audio, Speech, and Language Processing*, 16(5), July 2008.

[416] D. Yu, L. Deng, Y. Gong, and A. Acero. A novel framework and training algorithm for variable-parameter hidden markov models. *IEEE Transactions on Audio, Speech and Language Processing*, 17(7):1348–1360, 2009.

[417] D. Yu, L. Deng, X. He, and A. Acero. Large-margin minimum classification error training: A theoretical risk minimization perspective. *Computer Speech and Language*, 22(4):415–429, October 2008.

[418] D. Yu, L. Deng, X. He, and X. Acero. Large-margin minimum classification error training for large-scale speech recognition tasks. In *Proceedings of International Conference on Acoustics Speech and Signal Processing (ICASSP)*. 2007.

[419] D. Yu, L. Deng, G. Li, and F. Seide. Discriminative pretraining of deep neural networks. *U.S. Patent Filing*, November 2011.

[420] D. Yu, L. Deng, P. Liu, J. Wu, Y. Gong, and A. Acero. Cross-lingual speech recognition under runtime resource constraints. In *Proceedings of International Conference on Acoustics Speech and Signal Processing (ICASSP)*. 2009b.

[421] D. Yu, L. Deng, and F. Seide. Large vocabulary speech recognition using deep tensor neural networks. In *Proceedings of Interspeech*. 2012c.

[422] D. Yu, L. Deng, and F. Seide. The deep tensor neural network with applications to large vocabulary speech recognition. *IEEE Transactions on Audio, Speech, and Language Processing*, 21(2):388–396, 2013.

[423] D. Yu, J.-Y. Li, and L. Deng. Calibration of confidence measures in speech recognition. *IEEE Transactions on Audio, Speech and Language*, 19:2461–2473, 2010.

[424] D. Yu, F. Seide, G. Li, and L. Deng. Exploiting sparseness in deep neural networks for large vocabulary speech recognition. In *Proceedings of International Conference on Acoustics Speech and Signal Processing (ICASSP)*. 2012.

[425] D. Yu and M. Seltzer. Improved bottleneck features using pre-trained deep neural networks. In *Proceedings of Interspeech*. 2011.

[426] D. Yu, M. Seltzer, J. Li, J.-T. Huang, and F. Seide. Feature learning in deep neural networks — studies on speech recognition. In *Proceedings of International Conference on Learning Representations (ICLR)*. 2013.

[427] D. Yu, S. Siniscalchi, L. Deng, and C. Lee. Boosting attribute and phone estimation accuracies with deep neural networks for detection-based speech recognition. In *Proceedings of International Conference on Acoustics Speech and Signal Processing (ICASSP)*. 2012.

[428] D. Yu, S. Wang, and L. Deng. Sequential labeling using deep-structured conditional random fields. *Journal of Selected Topics in Signal Processing*, 4:965–973, 2010.

[429] D. Yu, S. Wang, Z. Karam, and L. Deng. Language recognition using deep-structured conditional random fields. In *Proceedings of International Conference on Acoustics Speech and Signal Processing (ICASSP)*, pages 5030–5033. 2010.

[430] D. Yu, K. Yao, H. Su, G. Li, and F. Seide. KL-divergence regularized deep neural network adaptation for improved large vocabulary speech recognition. In *Proceedings of International Conference on Acoustics Speech and Signal Processing (ICASSP)*. 2013.

[431] K. Yu, M. Gales, and P. Woodland. Unsupervised adaptation with discriminative mapping transforms. *IEEE Transactions on Audio, Speech, and Language Processing*, 17(4):714–723, 2009.

[432] K. Yu, Y. Lin, and H. Lafferty. Learning image representations from the pixel level via hierarchical sparse coding. In *Proceedings Computer Vision and Pattern Recognition (CVPR)*. 2011.

[433] F. Zamora-Martínez, M. Castro-Bleda, and S. España-Boquera. Fast evaluation of connectionist language models. *International Conference on Artificial Neural Networks*, pages 144–151, 2009.

[434] M. Zeiler. Hierarchical convolutional deep learning in computer vision. Ph.D. Thesis, New York University, January 2014.

[435] M. Zeiler and R. Fergus. Stochastic pooling for regularization of deep convolutional neural networks. In *Proceedings of International Conference on Learning Representations (ICLR)*. 2013.

[436] M. Zeiler and R. Fergus. Visualizing and understanding convolutional networks. arXiv:1311.2901, pages 1–11, 2013.

[437] M. Zeiler, G. Taylor, and R. Fergus. Adaptive deconvolutional networks for mid and high level feature learning. In *Proceedings of International Conference on Computer vision (ICCV)*. 2011.

[438] H. Zen, M. Gales, J. F. Nankaku, and Y. K. Tokuda. Product of experts for statistical parametric speech synthesis. *IEEE Transactions on Audio, Speech, and Language Processing*, 20(3):794–805, March 2012.

[439] H. Zen, Y. Nankaku, and K. Tokuda. Continuous stochastic feature mapping based on trajectory HMMs. *IEEE Transactions on Audio, Speech, and Language Processings*, 19(2):417–430, February 2011.

[440] H. Zen, A. Senior, and M. Schuster. Statistical parametric speech synthesis using deep neural networks. In *Proceedings of International Conference on Acoustics Speech and Signal Processing (ICASSP)*, pages 7962–7966. 2013.

[441] X. Zhang, J. Trmal, D. Povey, and S. Khudanpur. Improving deep neural network acoustic models using generalized maxout networks. In *Proceedings of International Conference on Acoustics Speech and Signal Processing (ICASSP)*. 2014.

[442] X. Zhang and J. Wu. Deep belief networks based voice activity detection. *IEEE Transactions on Audio, Speech, and Language Processing*, 21(4):697–710, 2013.

[443] Z. Zhang, Z. Liu, M. Sinclair, A. Acero, L. Deng, J. Droppo, X. Huang, and Y. Zheng. Multi-sensory microphones for robust speech detection, enhancement and recognition. In *Proceedings of International Conference on Acoustics Speech and Signal Processing (ICASSP)*. 2004.

[444] Y. Zhao and B. Juang. Nonlinear compensation using the gauss-newton method for noise-robust speech recognition. *IEEE Transactions on Audio, Speech, and Language Processing*, 20(8):2191–2206, 2012.

[445] W. Zou, R. Socher, D. Cer, and C. Manning. Bilingual word embeddings for phrase-based machine translation. In *Proceedings of Empirical Methods in Natural Language Processing (EMNLP)*. 2013.

[446] G. Zweig and P. Nguyen. A segmental CRF approach to large vocabulary continuous speech recognition. In *Proceedings of the Automatic Speech Recognition and Understanding Workshop (ASRU)*. 2009.

CPSIA information can be obtained at www.ICGtesting.com
Printed in the USA
LVOW09s1258040215

425654LV00001B/49/P